100+ TECH IDEAS FOR TEACHING ENGLISH AND LANGUAGE ARTS: MAXIMIZE IPAD, MOBILE AND ONLINE APPS IN EVERY CLASSROOM

Fitzell, Susan Gingras

100+ Tech Ideas for Teaching English and Language Arts:

Maximize iPad, Mobile and Online Apps in Every Classroom ISBN: Pending

---

Copyright © 2015 by Susan Fitzell

All rights reserved. Printed in the United States of America. All material in this book, not specifically identified as being reprinted from another source, is copyright© 2012 by Cogent Catalyst Publications. You have permission to make copies of handouts for your own personal use.

You may not distribute, copy, or otherwise reproduce any of this book for sale, commercial, or non-commercial use without written permission from the author.

Titles by Susan Fitzell, M.Ed.:

RTI Strategies for Secondary Teachers

Special Needs in the General Classroom, Strategies That Make It Work

Paraprofessionals and Teachers Working Together

Umm Studying? What's That?: Learning Strategies for the Overwhelmed and Confused College and ~ High School Student

Co-teaching and Collaboration in the Classroom

Please Help Me With My Homework! Strategies for Parents and Caregivers

Transforming Anger to Personal Power: An Anger Management Curriculum for Grades 6 through 12

Free the Children: Conflict Education for Strong & Peaceful Minds

Use iPads and Other Cutting-Edge Technology to Strengthen Your Instruction

Memorization and Test Taking Strategies - Professional Development Training Video Set

Motivating Students to Choose Success

For access to your FREE collection of supplementary materials specific to this text and seminar, go to tech.susanfitzell.com

A PDF of the corresponding Seminar Presentation with several additional apps(poetry, writing prompts, book creation apps) is included in the free download!

To contact the author:

Susan Gingras Fitzell

PO Box 6182

Manchester, NH 03108-6182

Phone: 603-625-6087

Email: SFitzell@SusanFitzell.com

Keep up with Susan on the Web!
- Twitter: @susanfitzell
- Twitter: @TheHomeworkGuru
- Facebook.com/SusanFitzellF3
- Facebook.com/TheHomeworkGuru
- Facebook.com/HowToPreventBullying
- LinkedIn: www.linkedin.com/in/susanfitzell
- Pinterest.com/susanfitzell/
- www.scoop.it/u/susan-gingras-fitzell
- YouTube: SusanFitzell

www.CogentCatalyst.com -

Practical strategy books & products that support parents and educators.

100+ TECH IDEAS FOR TEACHING ENGLISH AND LANGUAGE ARTS

Best, New Strategies for Using iPads®, Mobile Devices and Other Cutting Edge Technology to Strengthen English/Language Arts Instruction Error! Bookmark not defined.

Why should you read this book? 1
    This Book Saves you Valuable Time 1
    Apps, Icons, and, Prices are Constantly Changing! 1
    New To Tech in the Classroom? Here's Your Challenge 1

IMPORTANT INFORMATION Before We Begin 2
    The Purpose of this Book is to Share Apps that Enhance Instruction in the English & Language Arts Classroom 2
    It Was Free Yesterday! Hey!... and the Icon Changed Again! 2

Icons Used In This Book 3

A Simple, Yet, Powerful, Motivation Tool 4
      Ideas for Using Timers in the English Classroom 6

Apps for Notetaking & Writing 7
      Ideas for Using Note Taking Apps in the Classroom 10
      Ideas for Using PDF Notes and Similar Apps in the Classroom 11

Organizing Research and Digital Files 13
    Evernote: 13
      Ideas for using Evernote in the Classroom 14

Apps that Support Reading and Develop Literacy Skillsets 16
    Audio Books: 16
      Ideas for Using Audio Books in the English Classroom 19
      Tip: Kindle or Nook vs. PDF Notes 20
    Imagine More than Book: Downloading Complete Audio Courses! 21
    Tech Tools for Using Audio in the Classroom 22
    Tools for Determining Reading Levels 23

Content Curation for Higher Level Critical Thinking 24
      Ideas for Using Content Curation in the English Classroom 28

Targeted Applications for Specific English & L.A. Standards 29

Using Technology to Foster Recall and Recognition 31
    Use "Meaningful" Color: 31

Fostering Understanding Through Video Clips 32
    Apps that Capture and Play Video: 32
    Web Resources for Video to Use in Language Arts Lesson Plans 36

Tip: Search for Videos with Google Videos..........37
Ideas for Using Video in the Language Arts/English Classroom..........38

Using Music to Enhance Recall and Recognition..........39
Ideas for Using Music to Enhance Recall in the English Classroom..........40
Tip: Carry a Music Player on your Flash Drive..........40

Drill and Practice..........41

Deepening Understanding Through Applying Skills & Concepts..........42
Graphic Notes:..........42
Ideas for Using Collage and Multipurpose Design Apps:..........43
Cognitive Maps A.K.A. Mind Mapping..........44
Ideas for Using Digital Mind-mapping Tools:..........46
Ideas for Using Lino in the English Classroom..........47

Speech to Text: Is It Still Writing?..........48
Ideas for Having Fun with Siri..........49

Mechanics of Writing..........51
Citation Tools..........52
More Web Resources for the Mechanics of Writing..........54

Visual Cues - Scientifically Known as Nonlinguistic Representation..........56
The Brain Thinks in Pictures:..........56
Teaching Vocabulary with Visuals:..........56
Vocabulary Apps:..........57
Storyboards for Reading Comprehension:..........59
Create a Sequence Chart:..........60
Draw It So You Know It – Non-Linguistic Representation:..........61
More Ideas for Using Digital Image Apps:..........63
Ideas for Using Postcard Apps as Part of an English Lesson:..........64
Ideas for Using Haiku Deck as Part of an English Lesson:..........65
Ideas for Using Movie Makers in the English/Language Arts Curriculum:..........67
Ideas for Using Eyejot and Other Video Apps..........68
Ideas for Using Educreations:..........73

Research Apps..........75
Ideas for using Wolfram Alpha in English/Language arts:..........76

Cloud Storage and Teacher Tools..........78

Have Fun!..........80

Don't Forget to Download the Suppliemental Freebies!..........81

WHY SHOULD YOU READ THIS BOOK?

Descriptions of apps are freely available in the app store for your device, various review websites, other books, etc. So, you do NOT need to read this book for that information.

The reality is that there are thousands of apps out there. Of those thousands, thousands are not worth your time or money.

THIS BOOK SAVES YOU VALUABLE TIME

Consequently, my approach to the challenge of writing a book that provides teachers with value has been to download, test, and try iPad apps, portable apps, and a few apps that are cross-platform so that I could provide you with solutions that work and don't waste your time, or your students' time.

In the process, I will share some of the realities of shopping for apps, show you apps that work well, and tell you why I like certain apps and why I don't like others.

That said, because it is a book about apps, I will include app descriptions. Please understand that the VALUE is NOT in the app descriptions. The VALUE is that I've hand selected the best apps for you to use and that SAVES YOU VALUABLE TIME!

APPS, ICONS, AND, PRICES ARE CONSTANTLY CHANGING!

Another reality that we face is that as I write this book, months before it goes into publication, is that many of these apps and/or their icons will have changed by the time you read this book. There's only one change factor that presents a problem: Some apps disappear, become unavailable, or are renamed when purchased by a new developer. Sadly, I don't have a solution for that occasional problem.

NEW TO TECH IN THE CLASSROOM? HERE'S YOUR CHALLENGE

I challenge you to step out of your comfort zone and try some of these tools in your classroom or with your own child. For many students, this could make or break their success in school. At this very moment, rather than typing or writing, I am speaking what you are reading into Dragon Speak Professional, a voice to text program.

Yesterday, I mind-mapped an article using Popplet, then used Dragon Naturally Speaking to talk to each point in the mind-map. There's more than one way to organize ideas and write essays.

Our students may have thought provoking ideas to share. Brilliance need not require excellent writing skills to shine forth. Brilliance can present itself through multiple modes of communication. When we dismiss a child's brilliance because they cannot show what they know in a traditional education setting, for them, we deny the world that child's gifts.

IMPORTANT INFORMATION BEFORE WE BEGIN

THE PURPOSE OF THIS BOOK IS TO SHARE APPS THAT ENHANCE INSTRUCTION IN THE ENGLISH & LANGUAGE ARTS CLASSROOM

This book is based on the premise that you already know how to use the various devices, and how to set up device stations in your classroom.

If you are new to technology, in order to build your confidence in using the iPad, know that there are many tutorials, or user guides, available to assist you. In fact, you can find a user guide directly on your iPad by accessing Safari from your iPad toolbar, then the Bookmarks icon, then scroll down to the icon on the bottom that reads, "iPad User Guide."

Keep in mind that your students know more about device technology than you do. They are neither intimidated by technology, nor are they scared of taking risks with it. You must be of this mindset as well.

IT WAS FREE YESTERDAY! HEY!... AND THE ICON CHANGED AGAIN!

In my first edition of **Use iPads® and Other Cutting-Edge Technology to Strengthen Your Instruction,** I included pricing for all the apps. As I used the book in seminars on the topic, I realized that the pricing, as well as the icons, continuously changed. An app could be one price in the evening before a seminar and be a different price the next day. App and device updates might present a different image for an icon or completely different functionality for the app.

Consequently, I have decided to take the pricing out of this edition. It's too unreliable. Most of the apps I've chosen were free when I tested them. Also, icons may change, and iPad-only apps may be available on all devices by the time you read this book. So, even if I don't have an app designated as available for Android, search the Google Play store just in case it's now available.

Portable apps are one exception to the pricing dilemma. Because they are open source, they are consistently free. Some developers may ask for donations, however, these are free apps that work on a PC from a USB flash drive. That said, most can also be loaded directly onto to the PC if you like. I use Greenshot and VLC Player so often that it makes more sense to just have them loaded on my laptop.

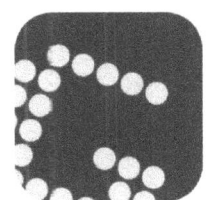

ICONS USED IN THIS BOOK

 **iPad**

 **Android**

 **PC**

 **WWW**

 **Portable App**

These icons are used in this book to denote apps that are available (as of the time of this printing) for the iPad, Android, or PC.

Apps denoted as WWW will work in almost any browser. The exception to this rule is that browsers on tablets and phones may not display certain www apps.

For example, Java browser apps will not display on all devices. Yet, they perform fine on a MAC or a PC.

Also, Portable apps are open source programs that live on a flash drive and can be taken with you to use on any PC. This is fabulous if you want drawing programs, graphic organizers, etc. available to you anywhere you go where there is a PC available

Consider a student who needs the ability to have text read aloud. If the student has the portable app with them in the computer lab, public library, or even at home, they may have the software to read text aloud. One example of such software is Balabolka.

A SIMPLE, YET, POWERFUL, MOTIVATION TOOL

Something amazing happens when a teacher introduces a timer into the classroom process. Students take notice and are often motivated at the idea of racing the clock.

Although a few students might be stressed by the use of a timer, the majority of students seem to enjoy the challenge, especially when a stopwatch timer is being used. If we think about it, it is really not that odd that today's generation would enjoy the race. Most of our students have grown up playing video games, many of which incorporate a "beat the clock" function. So this is a game they have been playing all their lives.

To download free supplements on motivating students, go to motivate.susanfitzell.com

Timer+ Touch HD

This is my favorite iPad timer app. It's so easy to use. Touch the app and move your fingers in a clockwise direction to set the time. The timer is shown in different colors for:
- Minutes in RED
- Hours in BLUE
- Seconds in GREEN

PC Chrono

PC Chrono has a timer, alarm, stopwatch, and countdown timer. It is an all-in-one portable app that is easy to configure and navigate. It is a very basic app that comes with limited features, but functions well.

PC Chrono
Use as a timer

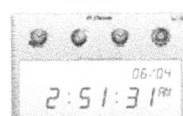

Online-Stopwatch

You will need to be online to use the browser app at Online-Stopwatch.com. It has a variety of different timer options but you must have an internet connection and your computer must be flash-enabled in order for you to use it.

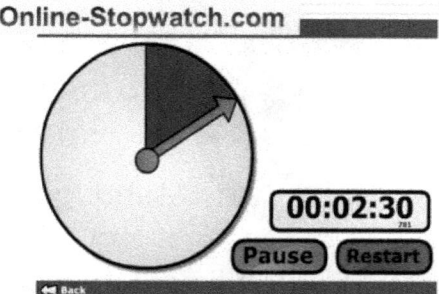

Hybrid Stopwatch and Timer:

Stopwatch and countdown timer with analog/digital display.

Music Clips & Ring Tones

Use instrumental music clips or cell phone ring tones as timers:

- Sort them by how long they take to play and use them as auditory timers for students during transitions or non-reading activities - the music will cue students in to the 'time' and keep them hopping.
- Avoid playing music while students are reading.

IDEAS FOR USING TIMERS IN THE ENGLISH CLASSROOM

1. Signal the beginning or the end of an activity.
2. Encourage students to stay focused on their task instead of wasting time watching the clock.
3. Nurture students' ability to take responsibility for effective time management.
4. Allow students to choose the timer format and sound to bring a bit of fun into the lesson and also to promote their sense of ownership of the process.
5. Ask a question and direct small groups to think of a possible answer. Use a one minute countdown timer to signal when time is up.
6. Allow a specific amount of time for students to plan their writing activity. During that time, they might use a linear outline or a graphic organizer. Tell them they are only allowed to do very specific prep work during that time. Too many students skip the prep work and jump right into the writing without the forethought necessary for quality piece of work.
7. Write a one syllable adjective on the board. Give small groups of students two minutes to come up with as many multisyllabic replacements of that adjective as they can think of, or find, using a digital thesaurus or a good old-fashioned paper-bound source. Be sure that the groups are mixed-ability so that no one group has an intellectual advantage.
8. What's your idea for using timers in the classroom? If you have a great idea, email me your idea and I will add it to my published material, citing you as the author.

APPS FOR NOTETAKING & WRITING

Apps for Notetaking – Whether Students can Write or Not!

Some learners are naturally good at taking notes. Yet, some students just can't seem to figure out what is important and what is not, what to write and what to leave out. And then, there are students who simply can't listen and write at the same time. So, to have those students take notes while the teacher is talking is counter productive.

Copying large amounts of notes during class time is also not the best use of a teachers' instructional time. There is so much to do in so little time that the time spent waiting for students to finish copying is a waste of valuable time.

So, consider how these note-taking apps might simplify and speed up the note-taking process as well as increase the accuracy of notes.

The iPad provides app choices for note-taking through the following inputs:

- Audio
- Handwriting/Drawing
- Text

One of the best features included in most of these apps is the ability to highlight, illustrate, and add pictures to notes. When students can accent their notes and their writing in different colors and make key words, letters, and other critical information stand out by writing it, bolding it, or underlining it in a different color, they will remember it better.

Adding visuals via photos and hand-drawn figures or mind maps takes the note-taking process up another level.

> *Non-tech strategy that supports students with visual spatial difficulties: Download your own printable Line Sheets to put under unlined paper to support neat, organized printing or writing, go to motivate.susanfitzell.com.*

Phraseology – A fabulous writing app that also determines reading level

Phraseology is a simple, distraction-free text writer that is incredibly powerful. I'm excited about the app because it allows the writer to draft paragraphs and move them around the document afterwards with the ease of drag and drop.

It also provides the writer with the reading level of their writing. Consider having students use complex sentence structure and multi-syllabic words to raise the reading level of what they are writing. It becomes a game and is a practical reality check for students to self-assess their writing.

*** Sometimes, with apps that save to the cloud, writing may be lost after an update. To be safe, copy and paste drafts and completed articles to another word processor, then save. ***

PHRASEOLOGY: WRITING ACCELERATOR
STUDENTS GET IMMEDIATE FEEDBACK ON WORD USE

AudioNote: - Let them record their notes so they can replay them!

This works like the LiveScribe pen. It can type or write text while recording, then play back notes with audio tagged to the notes. It will record as a student types and play back from specific words. The key is that the student MUST be within 10 feet of the speaker.

Check the law in your state regarding "right to know" for recording peoples' conversation before using recording devices in the classroom.

PenUltimate: For students and adults who prefer to write their notes

PenUltimate writes and prints better than all the other writing apps I tested.

PenUltimate now syncs with Evernote and that is a huge bonus for Evernote users.

- It syncs handwritten notes across devices.
- Handwritten notes are searchable.
- You must create an account to sync with Evernote. That may pose a logistical challenge in some school situations.
- Landscape is essentially unavailable with the exception of "Lined Landscape" paper in the Writing set of papers that is free for download.
- There is no recording option.
- There is no typing option. Penultimate is the digital version of handwriting on paper.

Notability: Now that this app records as well as organizes – it's a powerful productivity tool!

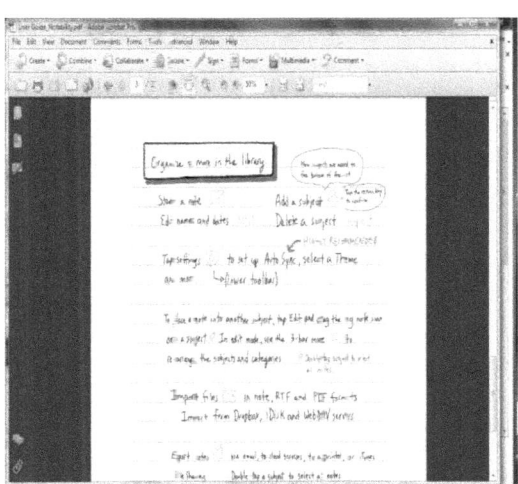

A note-taking app that is very popular and user-friendly. It includes a user guide to help you learn the app. It is a simple-looking app without a lot of visual images, but works very efficiently.

One of the things I liked about Notability's user guide is that it is short, clear, and to the point (about three pages). A key feature is the ability to annotate on top of content you have added.

It's great to use in the classroom to create tests or class notes and add to them directly from the screen as you're presenting the notes to the class.

For multiple how-to videos, type "Notability" in your YouTube search box.

IDEAS FOR USING NOTE-TAKING APPS IN THE CLASSROOM

1. Have students who struggle to take accurate notes use AudioNote so that what they miss in notes they can review with the audio recording.
2. Encourage students to capture the essence and details of a lesson by alternating between typing notes, recording the speaker - as well as their own thoughts - and illustrating their notes with the drawing tools.
3. Periodically, while students are taking notes, stop and ask a question. Have students record the question and the response using the audio recorder.
4. Use the note-taking app recorder to record an interview. Teach students to type/write key points and trust the audio capture to support a follow-on assignment.

PDF-Notes for iPad: Let's mark up that handout on our device!

PDF-Notes is an essential app for your iPad device that allows you to easily manage all your PDF files. The app offers the best finger writing experience with various pens, erasers, and highlighters. The user can easily import or export any PDF file via email, Dropbox, Safari, or iTunes.

It is a great app for both annotating a PDF file and writing simple notes. You can also save and share your files with others.

IDEAS FOR USING PDF NOTES AND SIMILAR APPS IN THE CLASSROOM

1. Rather than print handouts for students to read on paper, email students PDF versions of the lesson research, instructional notes, directions, or handouts. Students can then use PDF-Notes, or other similar apps, to highlight, illustrate, record comments, record questions etc. on the PDF document.

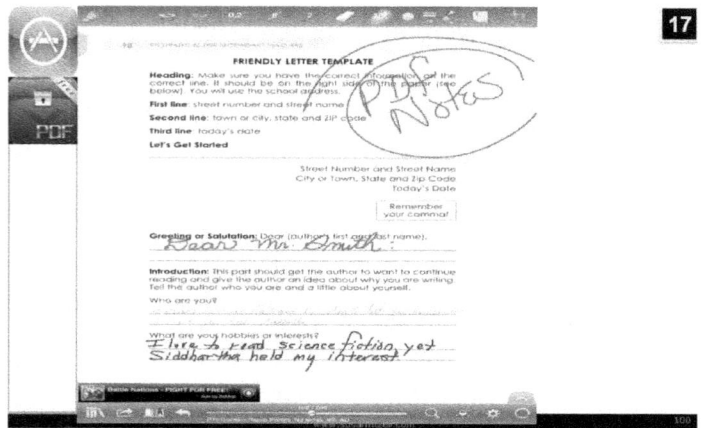

2. Digital books in PDF format can be read with PDF-Notes and highlighted or annotated.

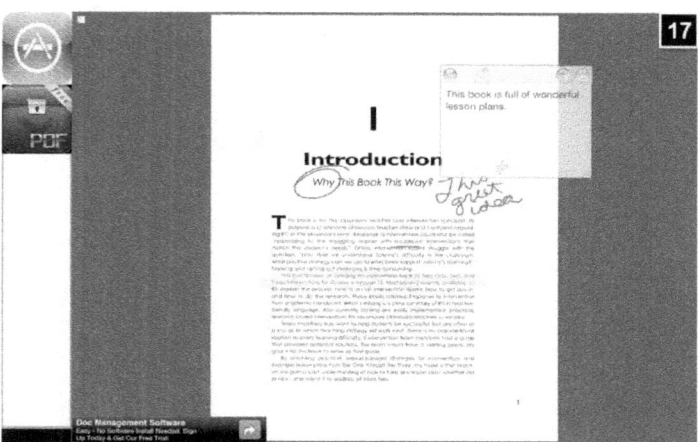

3. Use PDF-Notes as a tool where students may work with partners or in small groups to dissect written material. Documents may be shared through Dropbox, Google Drive, or your school's cloud or network storage.

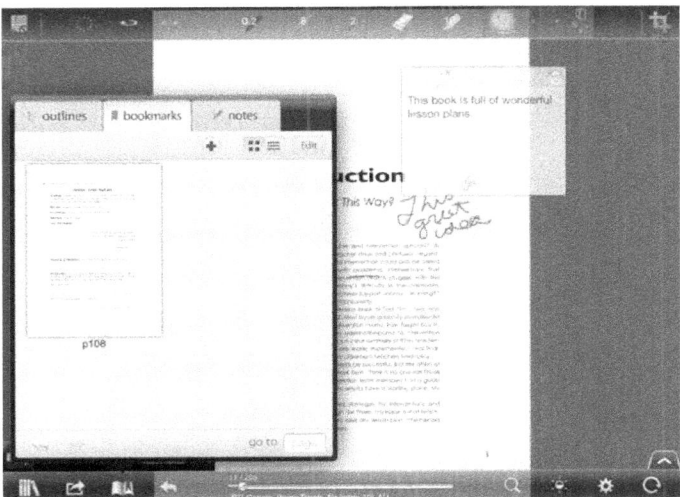

4. Create bookshelves of texts for students to access from the digital library.

5. Import a digital copy of a handout. Students can fill out the handout in PDF-Notes and email it, or export it, to return it to the teacher.

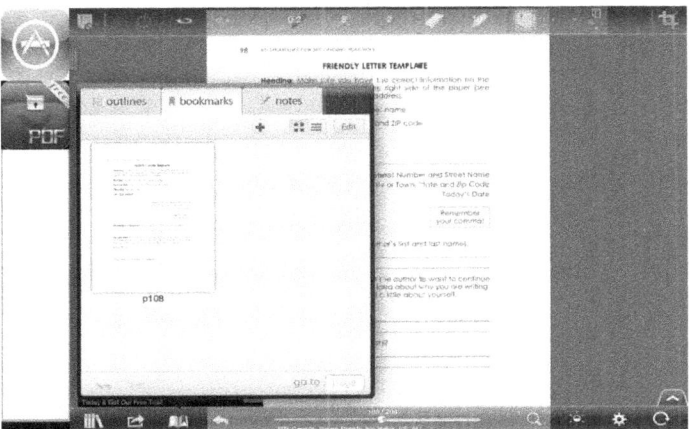

ORGANIZING RESEARCH AND DIGITAL FILES

EVERNOTE:

Using Evernote is like having an incredibly organized file cabinet with folders for every assignment, project, idea, interaction, or anything that you want to keep together in a safe place. Evernote is different than Dropbox in that Dropbox is like an external hard drive in the cloud. Dropbox holds files. Dropbox does not organize information or duplicate a file cabinet, unless the user somehow pulled that organization together. Google Drive is similar to Dropbox in that it holds files, however, its main advantage is that it allows multiple people to collaborate on projects, in real-time, from anywhere in the world.

Everything in Evernote is searchable, even this photograph of a whiteboard taken with the teacher's phone.

One can browse the web and find an interesting website that relates to research being done for an article or paper then click on the plug-in icon in the browser that will send that webpage directly to the file folder of choice in Evernote.

Because Evernote is cross-platform, anything stored in Evernote is visible from all your devices. It's like Microsoft Onenote on steroids. The only negative that I find in using Evernote as opposed to Microsoft Onenote is that if I want to add drawings or colors or visuals, I must do those in Skitch, which is a separate app unavailable on the PC. Skitch then syncs the newly created files with Evernote.

IDEAS FOR USING EVERNOTE IN THE CLASSROOM

1. Use Evernote to complete research for a project and share that research with the teacher.
2. Add articles by adding the "clipper" tool to your web browser.
3. As students read articles, they can add their own thoughts and ideas in Evernote, either in writing or via audio files.
4. When students take notes in class, they can use Evernote and share those notes with other students.
5. Students can add photographs taken with their devices to represent vocabulary, a setting in a play or novel, people they think might "look like" a character in a novel, etc.
6. Students use Evernote to create portfolios in the classroom. Handwritten notes can be scanned or captured with Evernote's camera function and added to their portfolio.
7. Teachers can use Evernote for researching and organizing their notes with the bonus of being able to search all notes in one place.
8. Teachers can create a notebook for each student and invite students to the notebook. Then teachers can comment on student notes.
9. Teachers can create one shared notebook for student assignments. Invite each student to that one notebook and list the days' work as well as the homework.
10. Teachers add digital versions of literature covered in class. Save the links in the notes and share them with students.
11. Post a public link to the notebook on the school website so that parents can join the notebook from the public link and support students with assignments.
12. Evernote's reminder function can support students to complete assignments on schedule.

For example: You surf the web and find a current event or editorial article related to the current lesson objective. You clip the article, include a video you found online related to the topic, include a picture gallery related to the report and add a corresponding worksheet that you will use in class. Add references to vocabulary in the article to the same folder. You then share an alert about this folder and assignment via Twitter or e-mail.

You can also install the "Clearly" extension for Evernote which eliminates ads, distracters, and web items to which you do not want students exposed.

Evernote Clearly lets you read your favorite content on the Web distraction free. A simple to use browser add-on for web browsers, Evernote Clearly removes unwanted advertisements, navigation bars, and other distractions with a single click, leaving only the content you care about in a clean, easy to read format. The text-to-speech feature in Evernote Clearly (for Chrome and Opera) even gives you the choice to have content read back to you when your attention needs to be elsewhere.

With Evernote Clearly you can also quickly clip content directly into your Evernote account to view anywhere you have Evernote installed, giving you an uninterrupted reading experience.
http://evernote.com/clearly/guide/

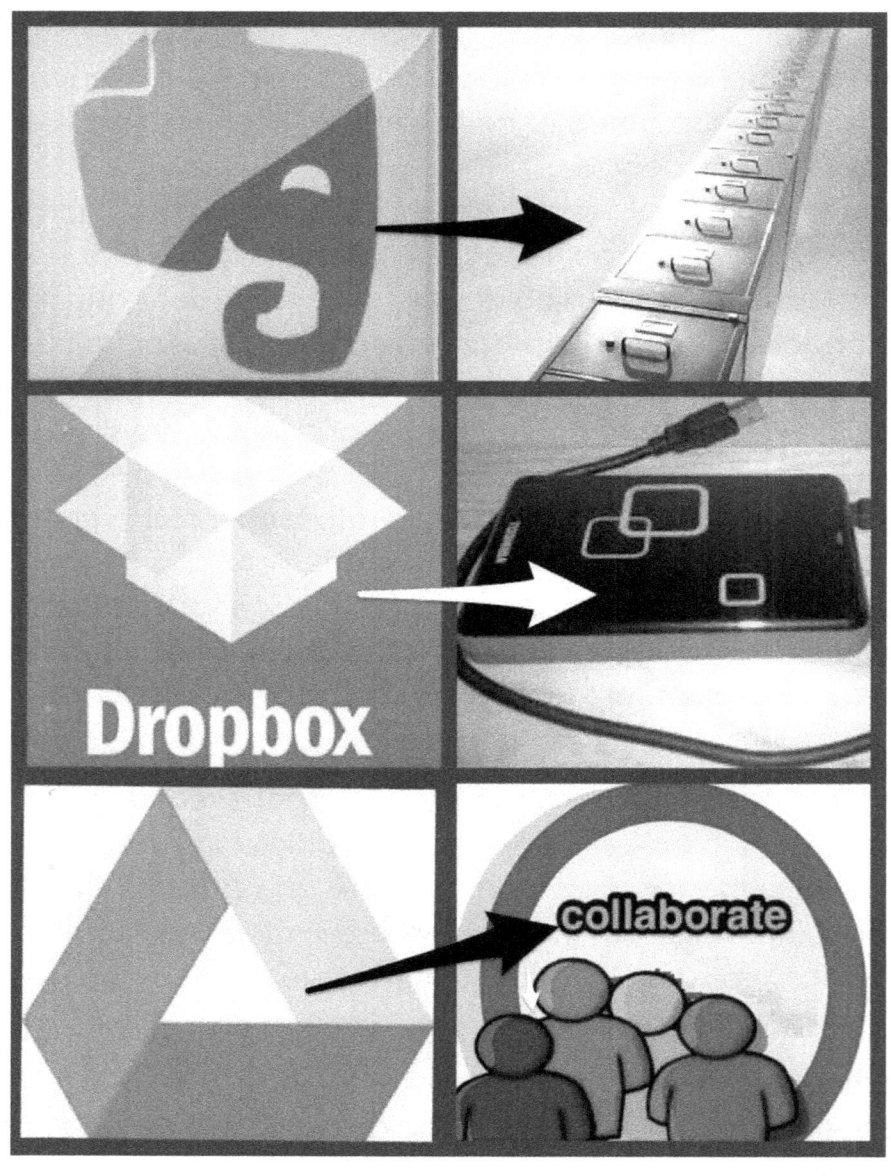

Visual Analogy to Understand These Apps

APPS THAT SUPPORT READING AND DEVELOP LITERACY SKILLSETS

Reading has always been fundamental to learning. Readers do not become readers unless it is taught and modeled to them. So, read to youth often. Reading will

- expose them to higher levels of vocabulary.
- model the importance of looking up and defining unknown words.
- raise their instructional level.

According to the National Center for Education Statistics (NCES), a division of the U.S. Department of Education, children who are read to at home enjoy a substantial advantage over children who are not.

- 26% of children who were read to three or four times in the last week by a family member recognized all letters of the alphabet. This is compared to 14% of children who were read to less frequently.

The NCES also reported that children who were read to frequently are also more likely to

- count to 20 or higher than those who were not (60% vs. 44%).
- write their own names (54% vs. 40%).
- read or pretend to read (77% vs. 57%).

According to NCES, only 53% of children ages three to five were read to daily by a family member (1999). Children in families with incomes below the poverty line are less likely to be read aloud every day than are children in families with incomes at or above poverty.

The more types of reading materials there are in the home, the higher students are in reading proficiency.

AUDIO BOOKS:

Listen to audio books as an alternative to TV, movies, and video games. Audio books combine important ingredients in creating a successful, lifelong reader. Audio books

- motivate students to read.
- allow students to enjoy a book at their interest level that might be above their reading level.
- allow slower readers to participate in class activities.

- provide a way to learn the patterns of language, learn expressions, and increase vocabulary.
- are good examples of fluent reading for children, young adults, and for people learning English as a second language.
- build the neural connections necessary for auditory processing skills. Auditory processing skills are required for literacy.
- improve listening skills.
- familiarizes students with the story (pre-reading) so that they can concentrate on the words when they read the text.
- build a reading scaffold -- broadening vocabularies, stretching attention spans, flexing thinking skills.

Audio book sources follow. Most of these sites charge for their products. Prices vary.

Recorded Book Rentals: <http://www.recordedbooks.com>

Recorded Books, LLC, is now a division of Haights Cross Communications, a premier educational and professional publisher dedicated to creating the finest books, audio products, periodicals, software and online services serving the following markets: K-12 supplemental education, public and school library publishing, audiobooks, and legal and medical publishing.

Books on Tape: <http://www.booksontape.com>

A division of Random House, Inc. Features unabridged audiobooks for libraries, schools and consumers.

The Audio Book Store: <http://www.theaudiobookstore.com/>

AudiobookStore.com is dedicated to providing the largest selections of audiobooks and formats available, in download, streaming & CD formats. With over 150,000 combined audiobook titles in our online catalog, our visitors can instantly download audiobooks to their favorite mp3 player, smartphone, tablet or computer, as well as enjoy other formats such as streaming audiobooks and CD audiobooks. In addition,

all titles in our catalog are fully compatible with (but not limited to) Apple devices including iPads, iPhones, and iPods, as well as all Android devices.

Learning Ally (Formerly Recording for the Blind & Dyslexic): http://www.learningally.org/

They have 75,000 unabridged books on tape. Students can get textbooks custom-recorded; ask for information.

LibriVox: http://librivox.org/

 Extensive collection of free audio books read by volunteers; the goal is to record every book in the public domain. LibriVox Audio Books provides free access to over 6,000 audio books. Each audio book can be streamed over the internet or downloaded for later use.

Learn out Loud: http://www.learnoutloud.com/

Over 30,000 of the best audio books, podcasts, free downloads, and videos you can learn from. "Apple iTunes now features a section of their store called iTunes University, which features free audio & video downloads from dozens of universities across the United States and around the world including Stanford, Duke, MIT, Arizona State, and more.

At LearnOutLoud.com we combed through these free resources to pick out the best lectures, courses, and audio & video programs that iTunes U offers."

Books Should Be Free: http://www.booksshouldbefree.com/

Free Public Domain Audiobooks & eBooks. Royalty free and public domain books and publications. Download books for iPhone, Android, Kindle & mp3 players

Free Books for iPad:

 23,469 classic books, with the ability to highlight, create notes, bookmarks and have dictionary support. To get the audio book functionality, you need to upgrade to the pro version.

IDEAS FOR USING AUDIO BOOKS IN THE ENGLISH CLASSROOM

1. Use these resources as another way to support those struggling readers, auditory learners, and English Language Learners.
2. Expose students to the patterns of language, it's rhythm and cadence.
3. Provide students the opportunity to learn expressions, and increase vocabulary by hearing them in context.
4. Use as a pre-reading activity. Play an excerpt and then ask students to predict what will happen next.
5. Bring a book to life, thereby inspiring, entertaining, and linking language and listening to the reading experience.

Amazon Kindle:

Access your Kindle even if you don't have your Kindle with you by downloading it onto your iPad or Android device. This app automatically synchronizes your last page read between devices with Amazon Whispersync. Adjust the text size, add bookmarks, and view the annotations you created on your Kindle. Get the best reading experience available.

Any PDF document can be e-mailed to (your name)@kindle.com or (device name)@Kindle.com and the PDF can be pulled into the Kindle app to read it.

Price: Free (the app is free and the only thing you would have to purchase would be any books purchased in the Amazon store.)

TIP: KINDLE OR NOOK VS. PDF NOTES

I would use PDF-Notes to read any document that I may want to notate, write on, etc. Amazon's iPad kindle reader does not have any note taking options.

Barnes & Noble NOOK:

Your favorite reading & entertainment -- on your NOOK, tablet, or smartphone

IMAGINE MORE THAN A BOOK: DOWNLOAD COMPLETE AUDIO COURSES!

Teachers, consider your massive workload. Imagine that you could find quality Audio courses on the state standards you are required to teach – done for you – or more realistically – available for you to use as a launching pad?

Students and Parents, consider that there will be times when a student needs reinforcement, or content explained differently. Imagine using the sources below for supplemental instruction.

The Great Courses: http://www.thegreatcourses.com/

 The course creation process that follows involves so much more than just placing a camera in the back of a classroom. We surround the world's greatest teachers with a team of experts who collaborate on crafting a customized and entertaining educational journey that's both comprehensive and fascinating.

My Audio School: http://www.myaudioschool.com/

 Children with Dyslexia are often compelled to work below grade-level in core content areas, such as history and science, simply because they are unable to read a grade-level text independently. Yet, most dyslexics are quite capable of working above grade level if the visual reading component is removed.

You must be at least 18 years of age to purchase a subscription to My Audio School. A subscription to My Audio School opens all their content.

iTunes University - for Audio Books:

 You can download a huge selection of audio books with iTunes University onto an iPad.

If there are choices when planning lessons with common core, check to see which common core book selections are available on iTunes University  before planning your lessons.

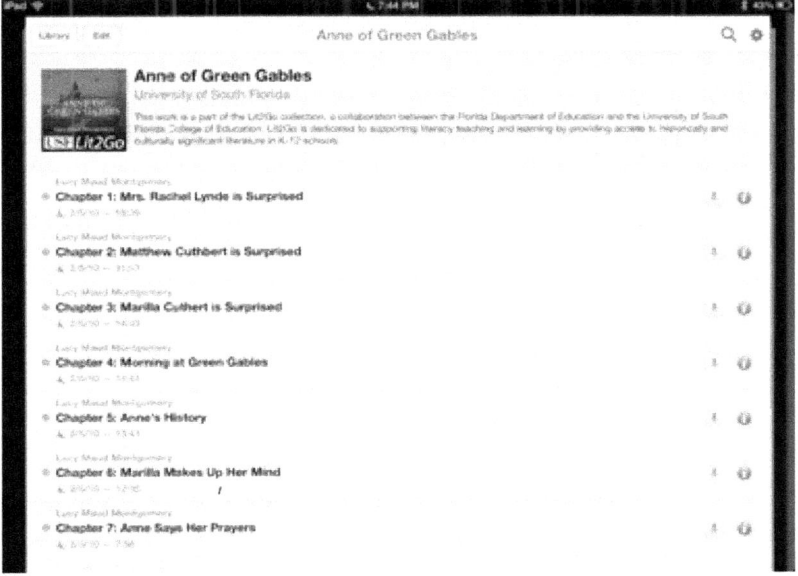

TECH TOOLS FOR USING AUDIO IN THE CLASSROOM

Balabolka:

A text-to-speech (TTS) application that uses the built-in voices of MS Windows to read text on-screen or any text passage you insert into the program. This is a great tool to support struggling readers, auditory learners, and English language learners.

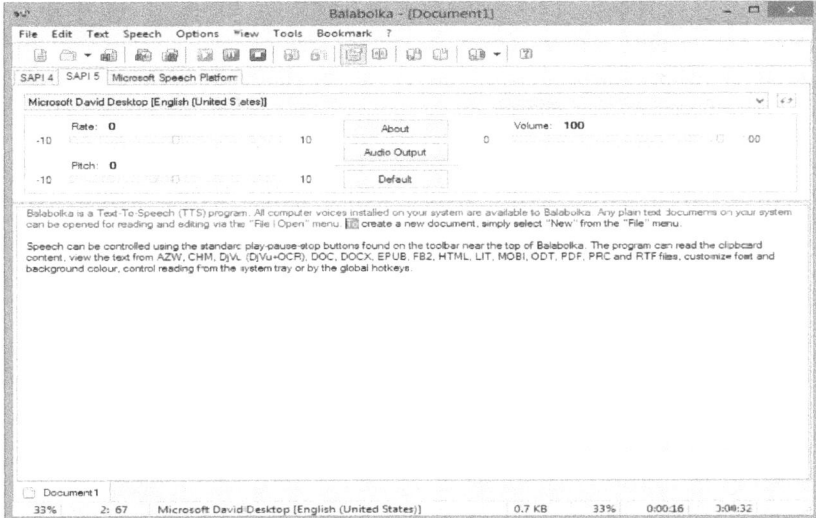

TOOLS FOR DETERMINING READING LEVELS

eReadability

 Pretty accurate app for determining the readability (how hard is it to decode this text) of print materials. I type in several sentence pairs and/or paragraphs from student texts and find out the readability and reading ease. These scores can't be compared to DRA or Lexile. They are drastically far apart in grade level results. Try it and you'll see what I mean.

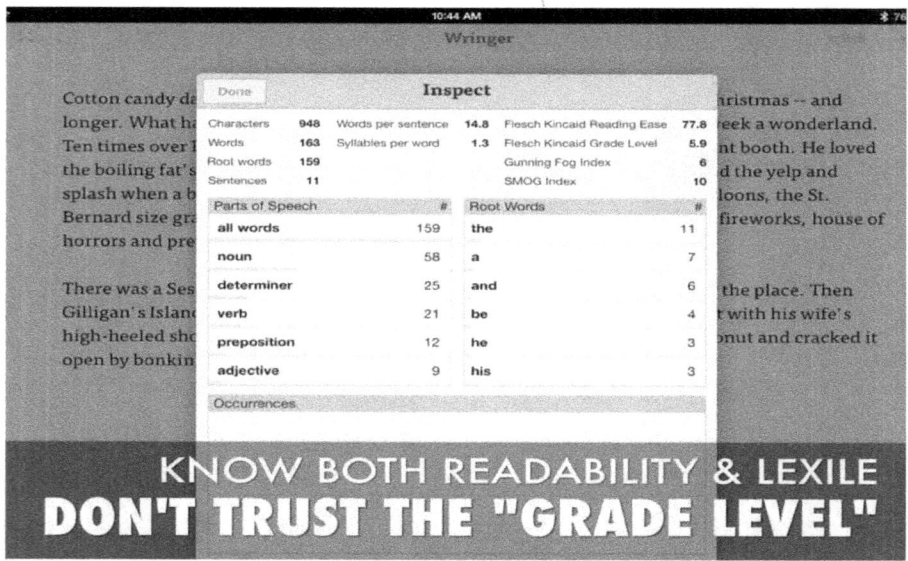

PHRASEOLOGY, FEATURED IN THE NOTE-TAKING AND WRITING SECTION, ALSO, PROVIDES THE READING LEVEL OF TEXT PASTED INTO THE APP.

CONTENT CURATION FOR HIGHER LEVEL CRITICAL THINKING

Pinterest is an example of a content curation site that has become enormously popular with teachers. Pinterest users spend hours collecting ideas, resources, items of interest, nostalgia and so on and so forth. Despite the popularity of Pinterest and other online resources for collecting and curating information, is "collecting" a worthwhile endeavor during valuable classroom time?

It seems the answer to that question lies in a teacher's ability to discern exactly what type of curation meets the educational objectives of the lesson. We have to look at the process as a value added activity that goes beyond collecting and classifying information under certain themes or topics.

To add true educational value, teachers need to consider how the collecting and curating process adds to the understanding of the lesson objective. How are the ideas connected? Will the curated content add to a depth of understanding for students studying a particular concept?

Will there be guidelines to decide what types of content will be collected? How will the information be organized? Will there be a unifying theme? What is the "why" behind what is collected? Is there a clear goal? How might curation be used to help students construct knowledge, as opposed to simply collecting reading material, videos, music, or other downloadable files? How do we support students in the curation effort and be comfortable knowing that we are using technology in a manner that is supported by evidence?

Nancy White addressed this issue in her article "Understanding Content Curation." She developed this excellent visual to support the use of curation beyond collecting.

Criteria	Collecting	Curating
Thinking Level	Classifying	Critical Thinking- Synthesis – Evaluation
Process	Not a lot of depth in the selection process; somewhat random	"cherry-picked" Reading, Synthesizing, Interpreting, Evaluating for Theme & Context; Disciplined, purposeful, continuous process of inquiry
Organization – How the resources are linked together	Thematic	Thematic AND Contextual – "real world" use, examples
Value	Meets a personal interest – value to collector. Quantity matters.	Meets a learning goal -value to collector AND learners. Quality matters
Audience	Not necessarily shared	Arranged, Annotated, & Published Somewhere – available to the general public – beyond the life of a particular "course" – Shared

[1]Graphic Source

---

[1] White, Nancy. "Understanding Content Curation." *Innovations in Education*. N.p., 07 Jul 2012. Web. 15 Sep. 2013. <http://d20innovation.d20blogs.org/page/2/>.

Flipboard

Flipboard is a social networking site that uses beautifully designed, magazine style, pages to curate information.

It's possible to read all your social networking sites through Flipboard as well as favorite news sources like the New York Times or Huffington Post.

What I didn't expect is the capability to view video from YouTube or other sites through Flipboard. No longer is a "magazine" solely a repository for text.

*NOTE: Currently, there are no parental controls on Flipboard because there is not an account system set up to allow for these types of monitors.

Scoop.it a.k.a. Read.it

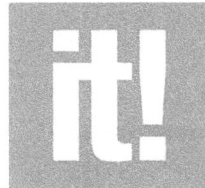

Scoop.it is one of the original curation sites. You can have up to five topics free of charge. The site provides suggestions for curating based your topics. It's easy to curate articles from suggestions offered. There's also a plug in for most browsers so that while surfing the web you can simply "scoop.it" by clicking on the browser plug-in icon to add the article to your Scoop.it category.

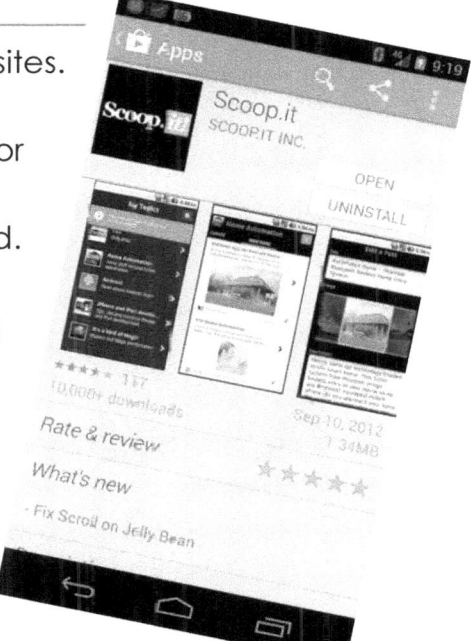

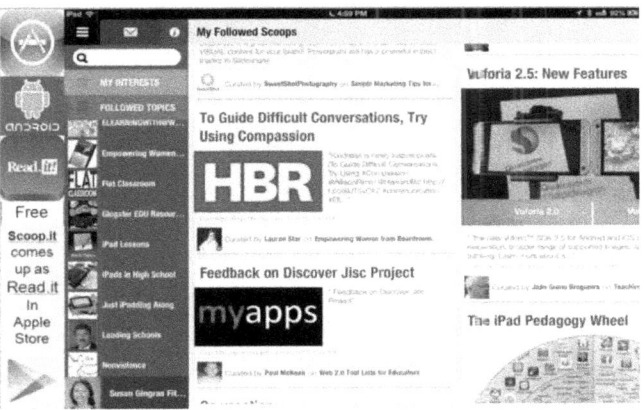

Take advantage of Susan's curated topics at Scoop.it/susanfitzell!

Pinterest:

Pinterest is a visual curation tool that people use to collect photos, project ideas, favorite products, articles (with a photo), as well as video. You can create and share collections (called "boards") of visual bookmarks (called "Pins") on any topic of interest. There are also "shared" boards that allow groups to pin items. It's possible for a teacher to have a "shared board" with students based on a topic being studied in class.

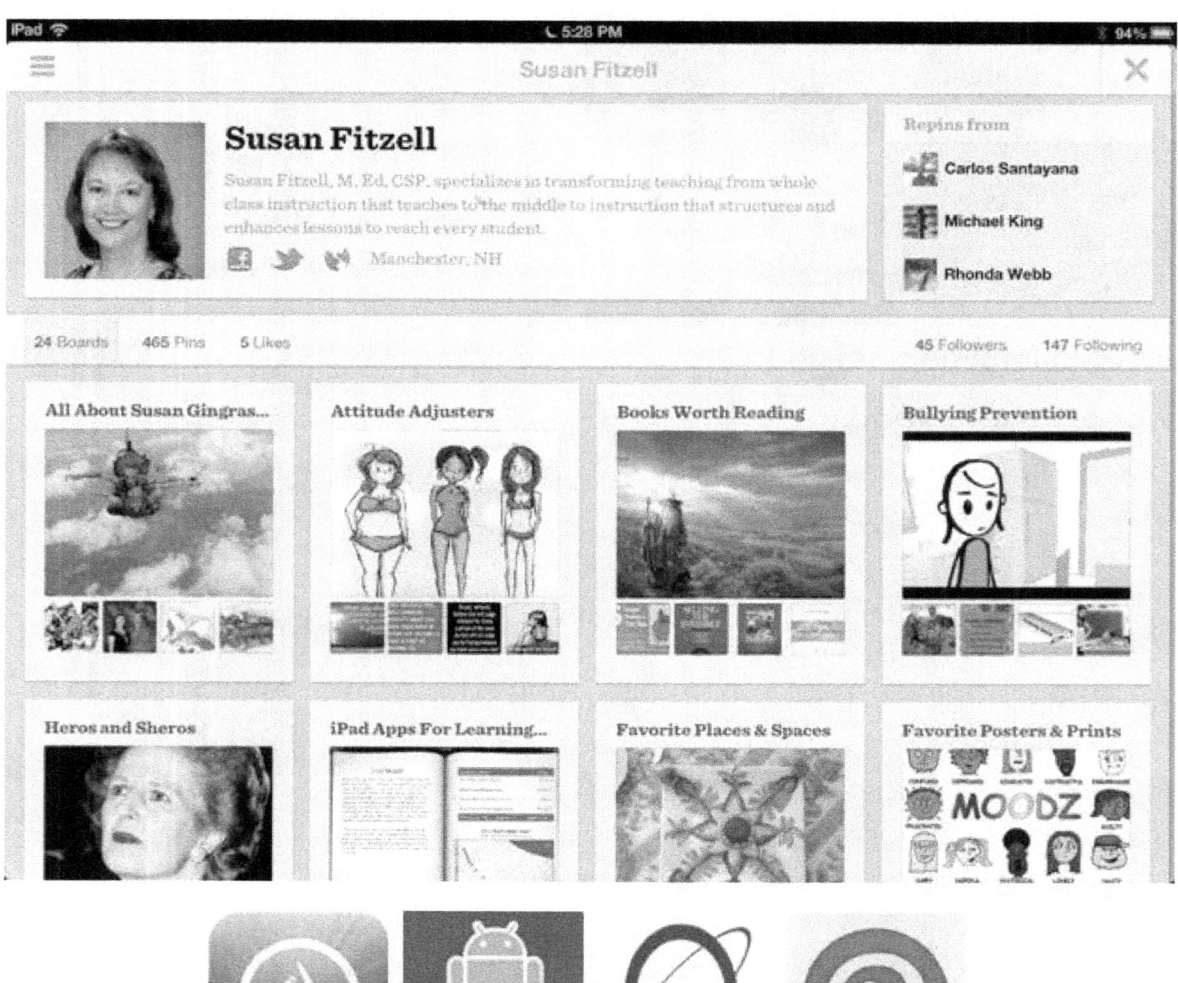

For additional apps to support learning in the classroom, go to http://www.pinterest.com/susanfitzell/apps-tech-for-learning-creating

IDEAS FOR USING CONTENT CURATION IN THE ENGLISH CLASSROOM

1. Prepare an assignment with materials and activities for your students and a curation tool such as Blendspace (formerly Edcanvas). Embed the canvas on a website, or possibly EdModo, for students to access during in-class projects or for outside assignments.
2. Have students curate articles and web resources to provide a "how to" resource. For example, students might measure influence in social media, impact of global climate change on the planet, or current issues related to a topic being studied in class. The key is that your goal is clearly aligned with the teaching objective and focused on curating in-depth information as opposed to just collecting a bunch of articles.
3. Create a board that includes a variety of video, articles, pictures, tutorials, and infographics that highlight content for language arts or English class curriculum objectives, such as verb tenses or figurative language.
4. Consider where students in your class are having difficulty and create a board that provides web-based options for learning and understanding the material better. These options can be accessed outside of school, thus providing double-dose access to the material.
5. Have students curate articles, pictures and video that spotlights a range of perspectives on an issue related to students' lives, or a piece of literature being read and discussed in the curriculum.
6. Create a rubric that guides students in evaluating resources they choose to curate. In this way, curating can become an assessment tool.
7. Aggregate, curate and create your own interactive textbook.
8. Create magazines with peers, as part of a class project, or to supplement the course text book.

TARGETED APPLICATIONS FOR SPECIFIC ENGLISH & L.A. STANDARDS

Shakespeare in Bits:

A new, exciting, multimedia approach to learning and teaching Shakespeare's plays. Shakespeare In Bits brings The Bard's most popular plays to life through magnificently animated re-enactments, full audio, and unabridged text in one comprehensive package.

As the video rolls, the text that aligns with the video highlights in red.

Price: Free for demo version. You get a sample of the famous balcony scene from Romeo & Juliet.

Open Source Shakespeare

"Open source" has two meanings: in the intelligence community, it means information that is published by normal distribution methods — say, a newspaper written in Urdu, or a television broadcast in Malaysia. In the computing world, it means a product whose source code is released freely, so other programmers can take portions of it for themselves, or revise and extend the original product. Prominent examples of open source software include the Linux operating system, the Firefox browser, and the Apache Web server, which runs about two-thirds of all public Web sites.

Open Source Shakespeare is open in both senses. The general public can use the site without paying money, or even registering for the site at all. Further, anyone is free to download and use any part of Open Source Shakespeare.

http://www.opensourceshakespeare.org/

English Idioms Illustrated:

This is an absolutely gorgeous app. I learned so much about the origin of many idioms from it. The free version gives you a good amount to work with. However, it's inexpensive enough for the entire package that a teacher might use it with an AppleTV or connect it to a projector for instruction of the entire class.

When students have difficulty understanding idioms, providing them with text explanations doesn't typically foster understanding or recall.

However, when presented with a visual depiction and explanation of the origins of the word, it will increase understanding and visual recall.

Use this app not only to help students understand idioms, but to also provide a model of illustration so that students can create their own visual idiom explanations.

USING TECHNOLOGY TO FOSTER RECALL AND RECOGNITION

USE "MEANINGFUL" COLOR:

Use color. Studies show that we remember color first and content second, so highlight or use colorful markers and pens to write vocabulary words and their definitions. Use different colors to make key words of a definition stand out and to help students remember the meaning of words.

Research shows that we learn better with color than with black and white text because color makes text unique, and the brain remembers what is unique. This is another reason to have students take their notes with colored fine tip markers or gel pens.

Skitch:

With Skitch, I can take a screen shot of anything on my iPad and take notes on the screenshot.

I've heard of some teachers using Skitch in their classroom to show a journey on a map by taking a screenshot from Google Earth and uploading it, then creating a diagram for students to view.

  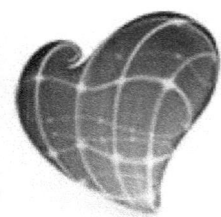

FOSTERING UNDERSTANDING THROUGH VIDEO CLIPS

Interweave your lesson with video clips that provide meaning and background for the topic being taught. Limit clips to 3-6 minutes.

APPS THAT CAPTURE AND PLAY VIDEO:

Video capture and video play have always proved challenging in the classroom. Given the uncertainty of buffering streaming video, blocked websites, and the inability to download or play certain file types, many teachers simply avoid video.

Avoiding video is the least desirable option.

Imagine that you are reading a story where the storyline takes place primarily in coastal waters and island habitats. Your students live on a Navajo reservation in the middle of Arizona. None of your students have had the benefit of ever seeing or experiencing an ocean. Some may not even have electricity or television at home, and consequently may not have seen movies or television programs set around coastal waters. If this sounds outlandish to you, I assure you that this situation exists as I have personally worked in Navajo schools as a presenter and consultant.

If we cannot take students to the ocean, we can at least show them video that allows them to experience the environment virtually. Students will be better equipped to understand the intricacies of a storyline if they understand the setting. So much of the literature students read in the classroom has no connection to their experiences or daily lives. Video can create that connection.

DamnVid:

A video downloader. While DamnVid can convert local video files, it can also download video streams from most video sharing websites. But what gives it the edge over other video downloaders and converters is that it downloads and converts at the same time, making the whole process much faster.

DamnVid has a built-in YouTube video search. Search and download directly from within the app!

VLC Player:

This is one of the most popular media players on any platform. Take your audio and video files, along with everything you need to play them on the go. You can place it on your USB flash drive, iPod, portable hard drive or a CD and use it on any computer or iPad. It is the only media player I use with one exception: Microsoft PowerPoint will only play video from inside the program if it is played through Windows Media Player. Consequently, I have that program on my computer. The only time I use it is when I am playing video through PowerPoint.

Price: Free

iTunes University - for video, audio, notes and more!

From the iTunes U app, students can watch a video or listen to audio lectures as well as download notes that are synchronized with the lecture.

Simply put, iTunes University is one of the best resources available on the Internet and if you never use any

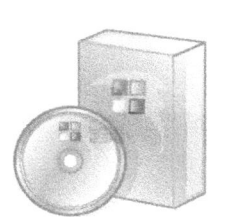

other Apple product, iTunes University, loaded on your PC, is worth the effort, simply for the abundance of free courses available to anyone with the desire to learn.

As a teaching application and a resource to support student understanding, it is unsurpassed in the quality of material available to choose from. English teachers might use iTunes University to supplement coursework in their current curriculum. Students might watch, read, listen to material on a topic they are learning in class on their own time for homework or for chunks of classroom time, as appropriate.

Teachers might upload their own course material to iTunes University, creating their own online course for students, or they might use a University or other grade level teachers' uploaded course.

Courses can be used to enhance the general curriculum or for extension activities for students who need to go beyond what is available in the classroom. The possibilities are endless and the resources are often free and of high quality.

Some examples of what's available:

The TED course on Media and Journalism provides some powerful video of media executives, journalists, and authors who have spoken on TED Talks. These videotaped speeches inspire thought, reflection, and awareness about journalism in the Western world.

This course is a free and powerful supplement to an English curriculum. What I find so exciting about using this vehicle as a method to promote knowledge and understanding is that it is presented in a format that is accessible to all ability levels. Students who struggle to read need not be left out of a powerful learning experience.

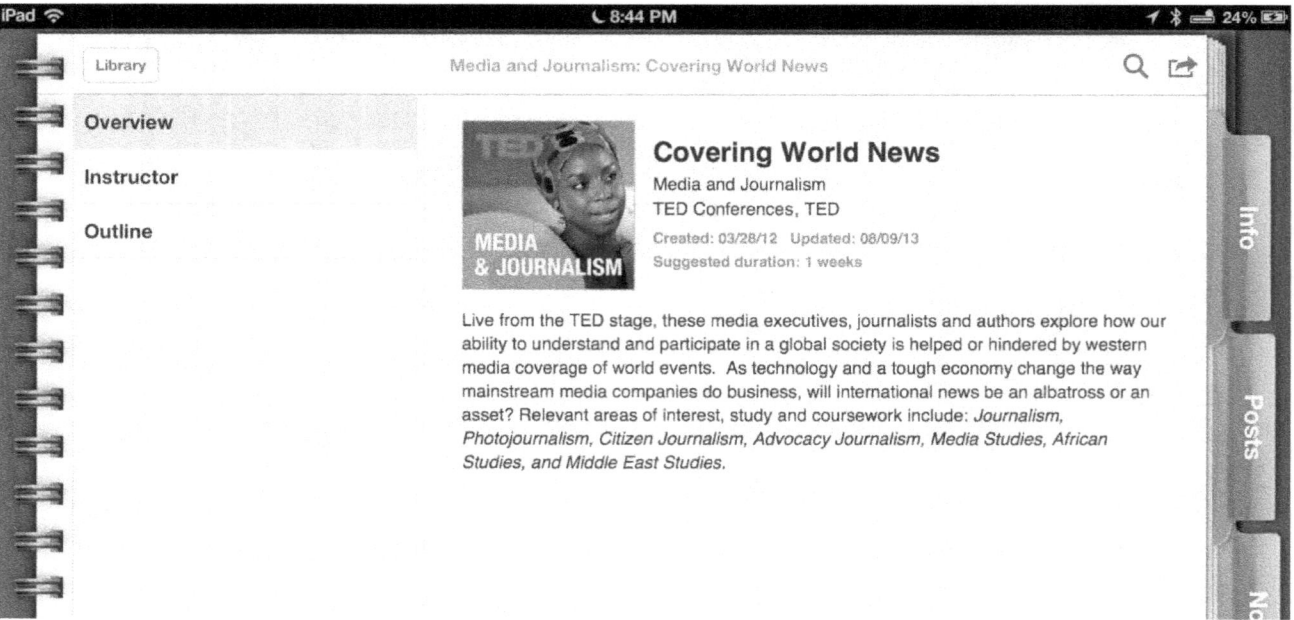

The next course example, Digital Storytelling, is a complete course that teaches students how to use digital storytelling via graphics with text, comics, narrated images, books, with or without audio, talking characters, animated stories, screen casting, and video.

Teachers do not need to spend hours finding the apps and websites required to create these digital learning experiences because all of those apps are linked to in the Materials section of the course. Also, the majority of the courses are free. Anything one might need in order to learn digital storytelling and generate product as an assessment of their learning is at the teachers' and students' fingertips with this free iTunes University course.

[Screenshot: iPad view of "Digital Storytelling" course overview by Tonia McMillan, Dawson Education Service. Created: 04/16/13, Updated: 09/03/13, Suggested duration: 1 weeks.

This is an resource/training session that focuses on the use of the iPad and Web2.0 sites that students and teachers can use to help in creating digital stories. Participants will learn the process of digital storytelling revolving around the CCSS in Language Arts and the integration of technology. All grade and content levels can benefit from this training course.

Many specific content standards can be addressed through digital stories. Here are *some* of the Common Core Standards that digital storytelling and digital stories address:

- CCSS.ELA-Literacy.CCRA.W.6 Use technology, including the Internet, to produce and publish writing and to interact and collaborate with others.
- CCSS.ELA-Literacy.CCRA.W.3 Write narratives to develop real or imagined experiences or events using effective technique, well-chosen details and well-structured event sequences.
- CCSS.ELA-Literacy.CCRA.SL.4 Present information, findings, and supporting evidence such that listeners can follow the line of reasoning and the organization, development, and style are appropriate to task, purpose, and audience.
- CCSS.ELA-Literacy.CCRA.SL.5 Make strategic use of digital media and visual displays of data to express information and enhance understanding of presentations.
- CCSS.ELA-Literacy.CCRA.SL.6 Adapt speech to a variety of contexts and communicative tasks, demonstrating command of formal English when indicated or appropriate.]

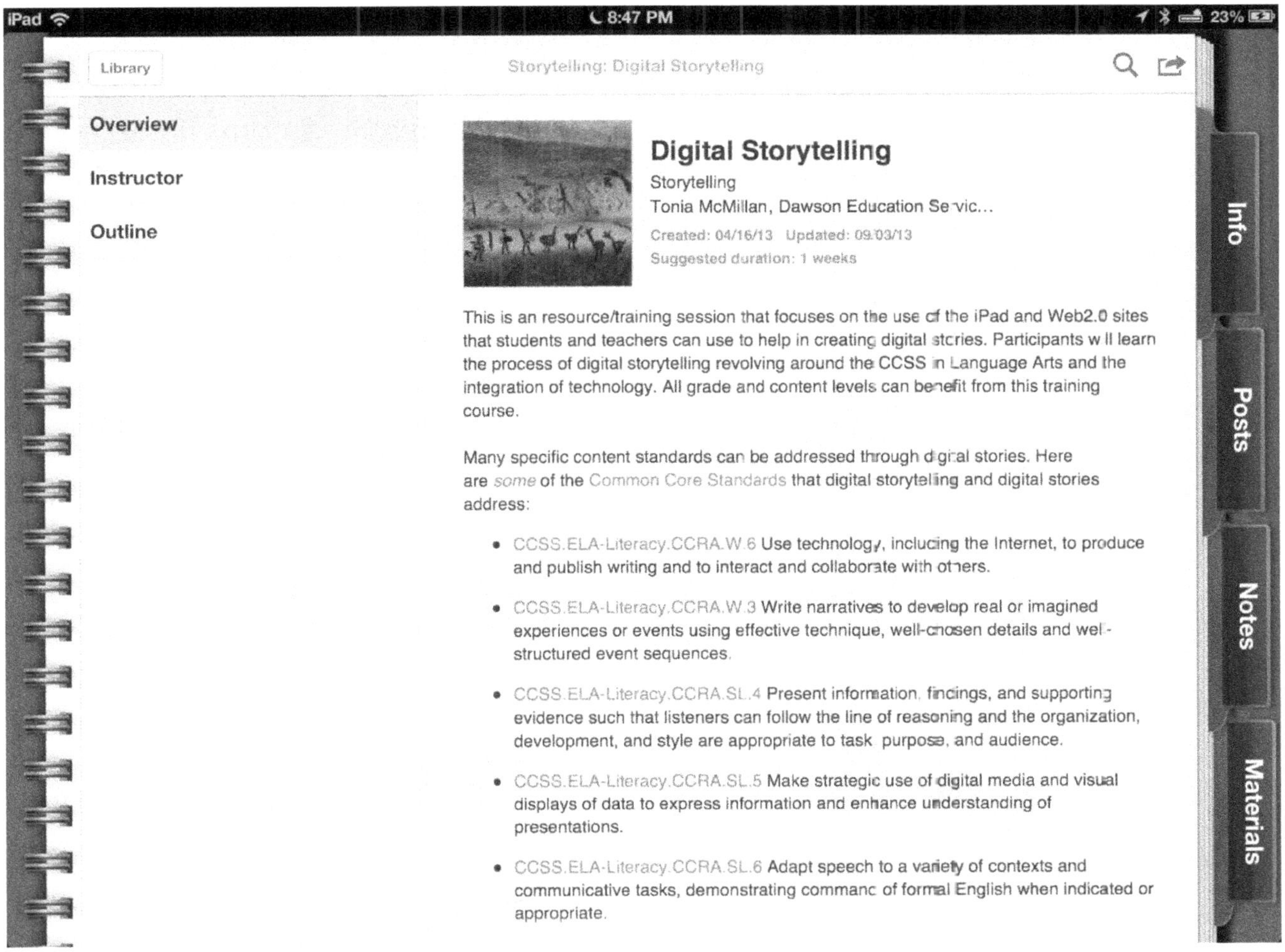

WEB RESOURCES FOR VIDEO TO USE IN LANGUAGE ARTS LESSON PLANS

Lessons on Movies - a site geared towards English as a Second Language, however, is just as valid for use in any Language Arts or English curriculum. Includes movie trailers and several activities related to the trailer to enhance language skills.

http://www.lessonsonmovies.com/

Breaking News English - another site geared towards English as a Second Language that creates lessons around odd "Breaking News" stories.

http://www.breakingnewsEnglish.com/

TIP: SEARCH FOR VIDEOS WITH GOOGLE VIDEOS

Rather than search for video to use in your classroom from Teachertube, Schooltube, Youtube, etc. Go to one source: Google Videos. Google Videos searches all three of those sites as well as university sites, Prezi.com, vimeo.com, showme.com, and countless other sources.

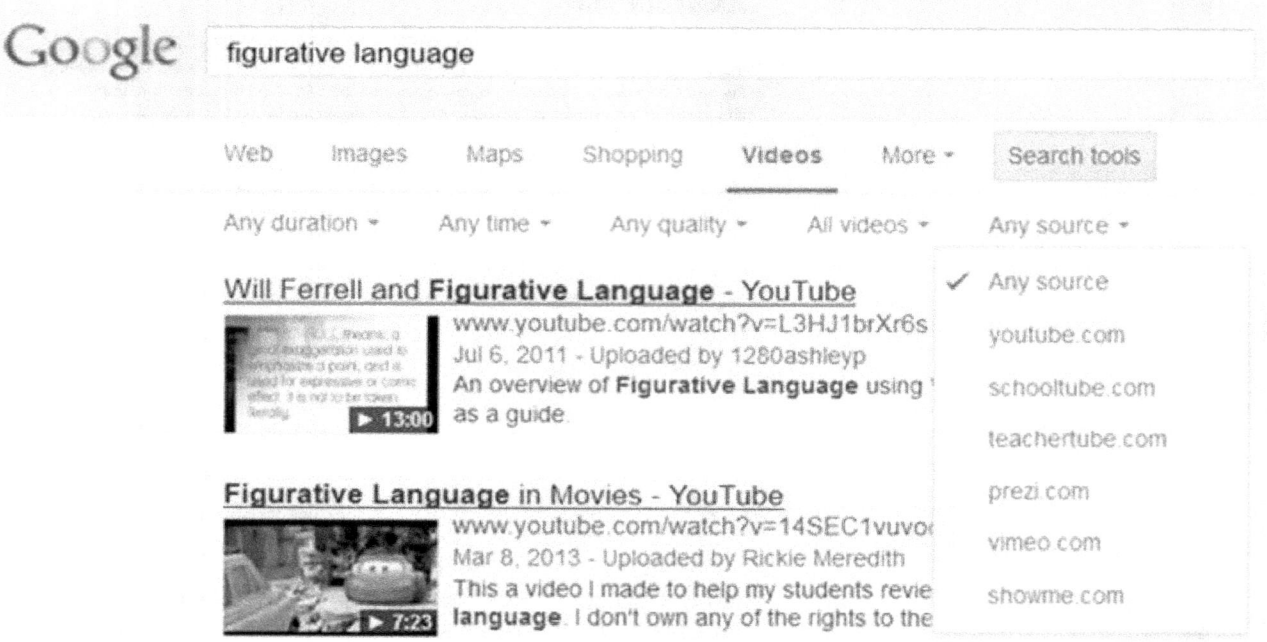

IDEAS FOR USING VIDEO IN THE LANGUAGE ARTS/ENGLISH CLASSROOM

1. Students watch a movie trailer and then discuss what aspects of the story cycle were utilized in the development of the trailer.
2. After watching a movie trailer, students work in small groups or pairs to identify the title, author, illustrator, setting, characters, and aspects of the plot featured in the movie trailer. Take it up a level and ask students to identify what emotional hooks were used to create a desire in the viewer to watch the movie.
3. After watching a movie clip, students work in pairs or small groups to create five questions about the movie for response from their peers. They must ensure that their questions reflect a sample of various levels of Blooms Taxonomy or Question Answer Response.
4. Students choose three to four videos on a piece of literature created by professional actors, other students, or educational sources and note the differences between the different video renditions. Which rendition did they prefer and why? Did students notice differences in perspective, bias, cultural attitudes, etc.
5. After watching a video clip, students brainstorm new vocabulary used in the clip and note the context and the meaning of the words, based on the clip.
6. Students write a press release for a movie being released soon. The movie does not need to be an "official" movie.
7. Use movie clips to teach:
   - characterization
   - grammar
   - drawing inferences
   - figurative language
   - vocabulary
   - persuasion
   - point of view
8. For homework, have students find video clips related to the topic taught during the day's lesson.

USING MUSIC TO ENHANCE RECALL AND RECOGNITION

Ever notice how you can easily remember the words to *The Itsy Bitsy Spider* and other songs you may not have sung since childhood? That's because the rhythm and rhyme of music helps you to remember the lyrics. This idea can be used to help students memorize vocabulary words by turning the words' definitions into song lyrics or by writing lyrics using sentences that put the words into an easy to understand context.

The deep part of our brain, the part that's more primitive, remembers music. Music is a powerful learning tool that is being used all over the country to increase long-term memory and boost state test scores.

One teacher taught the rules for long division using the tune and movement of *The Macarena*. When it came time for the math section of the state test, the test proctor knew exactly which kids were in this teacher's class because every now and then, one of them would stand up and start doing *The Macarena*. Then they would sit back down and begin to write.

Some kids won't sing 'silly' songs like algebraic equations to *God Bless America* or the quadratic formula to the tune of the *Notre Dame Fight Song*. In these cases, teachers use karaoke CDs or MP3s with current popular music and have their students rewrite the lyrics to fit the needs of the course material. Music is a powerful strategy. You can take anything you want students to learn, find any kind of rhyme or music they already know, and have them substitute the lyrics with material you are trying to teach.

Karaoke Music:

With the right equipment you can utilize music and put the talents of your students on display. It also makes for a fun day in music class. Or you can even use it with a project in History or English class.

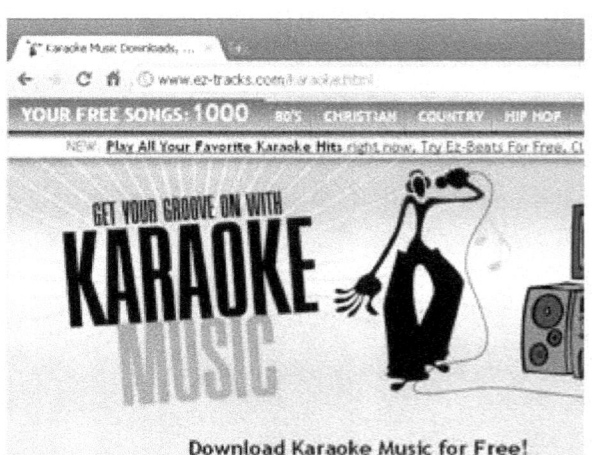

IDEAS FOR USING MUSIC TO ENHANCE RECALL IN THE ENGLISH CLASSROOM

1. Rewrite karaoke songs with information students need to learn.
2. Link old tunes with concepts: replace the words of a familiar song with information students must memorize.
3. Put the words to music!

TIP: CARRY A MUSIC PLAYER ON YOUR FLASH DRIVE

CoolPlayer:

 This portable app is an MP3 player with Freeform skin support. It features a playlist editor.

It is downloadable for Apple and Android users. It can also be downloaded to your hard drive to play on your computer

DRILL AND PRACTICE

SAT Question of the Day for iPad:

 From sat.collegeboard.org, this website and app give users a chance to tease their brain with real SAT questions and SAT preparation materials from the test maker. While there are books and other study materials available for a fee, the Question of the Day app is free.

SAT Word A Day AUDIO:

 Get a new SAT vocabulary word on your device every day.

There are several SAT prep apps available in Google Play.

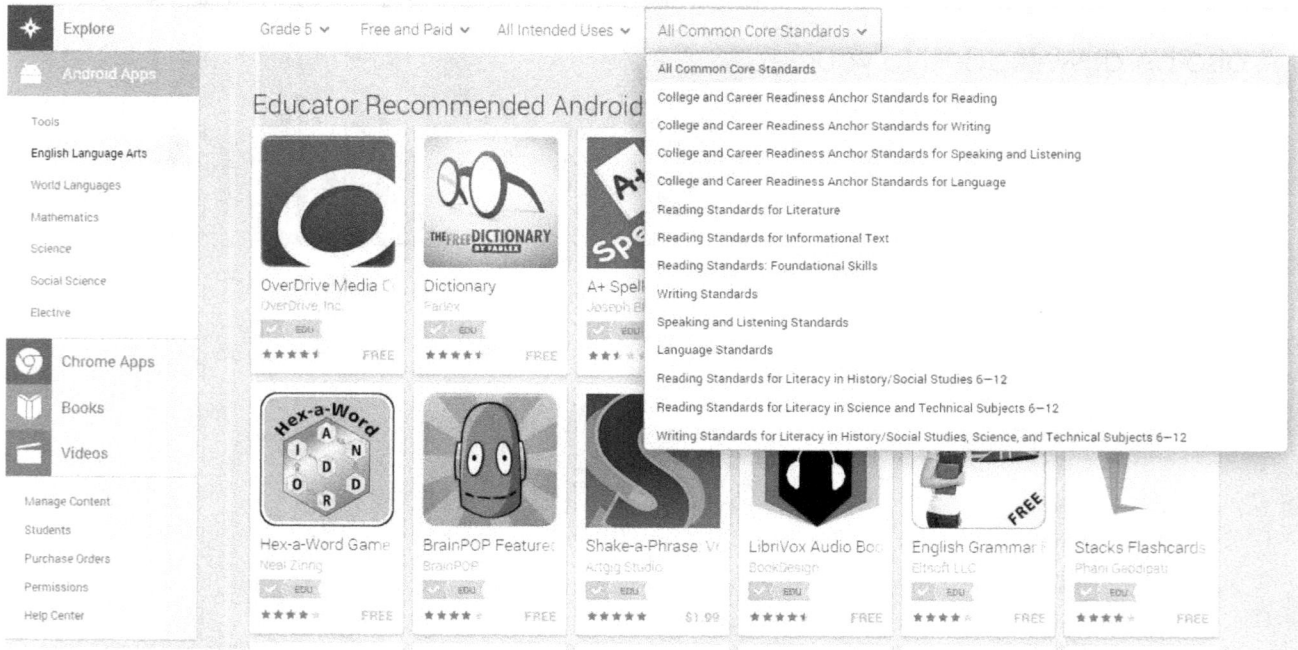

DEEPENING UNDERSTANDING THROUGH APPLYING SKILLS & CONCEPTS

GRAPHIC NOTES:

Graphic notes are a highly effective way to introduce a new concept or topic to students. To utilize the pictures in the textbook, or to find related pictures and provide students with the opportunity to

- preview the chapter or section of your reading.
- choose an image that is central to the topic.
- choose four areas of emphasis (may be subtopics).
- jot down key ideas under each heading.

Pic Collage:

  Pic Collage is an extremely easy to use collage maker. It can be used for any study or homework assignment where a collage would be appropriate.

RealWorld Paint:

This portable app allows you, or your students, to create an image from scratch or edit one of your own images. It also has many special effects, similar to Photoshop, like image blending, shadow effects, and retouching tools.

IDEAS FOR USING COLLAGE AND MULTIPURPOSE DESIGN APPS:

1. Ask students to demonstrate application of knowledge with prompts such as:
   - What examples can you find to_____?
   - Visually show how you would use _____.
   - Illustrate what approach you'd use to _____.
   - What images would you select to show how to apply _____?
2. Using a picture related to class reading; ask students to write a headline for the topic or issue that captures the most important aspect.
3. Look at an image; List 10 words or phrases about the image. Give extra points for multisyllabic words.

Take it up a level: Analyze:

- Use Skitch to show how you might design _____.
- Illustrate the differences between two characters' perception of an event or...

COGNITIVE MAPS A.K.A. MIND MAPPING[2]

Have students create a cognitive map with words and pictures to help them remember the sequence of events in a story or lesson. Using colored pencils and crayons, or even a computer, students can create a step-by-step map of the story or process, complete with descriptions of events and drawings or clip art to help them remember what happened.

Mind mapping is used to draw diagrams to show relationships between different ideas. They are a great visual tool to help students understand how ideas or tasks connect.

For many students, mind mapping the framework of an essay or story is much more effective than creating a linear outline.

Inspiration Software:

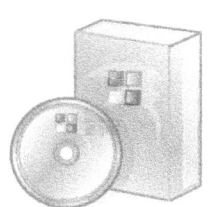

This is a desktop program that introduced an iPad app earlier this year. It promotes visual learning through the use of diagrams and cognitive mapping. It is a very useful tool to teach students the connection between different ideas or stories. This is a great tool to use in the classroom, especially for visually-driven students.

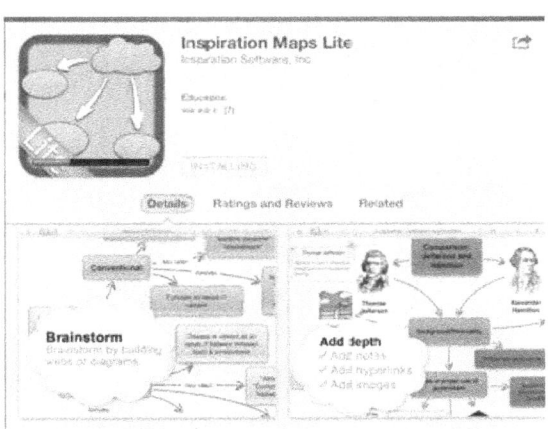

---

[2] Developing cognitive maps and using advance organizers also increases critical thinking skills. (Barba and Merchant 1990; Snapp and Glover 1990; Tierney, et al. 1989).

There are many graphic organizer apps for both the iPad and the Android as well as web-based apps and portable apps for creating mind maps. So why would you pay $9.99 for a mind mapping app when there are so many available for free? The answer, in my view, is simple: Inspiration software provides teachers and students with ready-made templates that can be used immediately to complete assignments.

For example, there are 18 ready-made English templates, complete with instructions for how to use them in the PC version of Inspiration software. Included are mind maps for literary comparison, literary conflict, persuasive essay, story triangle, and textual analysis. Add to that list 11 additional templates that could be used in the English classroom to foster thinking skills. For example, included in these thinking skills templates are analogy, comparison, problem solution, and opinion proof.

To my knowledge, no other mind mapping app or program includes the wide range of ready-made templates available to the student or teacher using Inspiration Software.

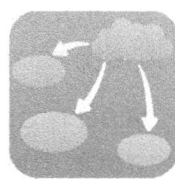

OmniGraffle:

Need to create a quick diagram, process chart, page layout, website wireframe, or graphic design? With OmniGraffle, your iPad touch screen is your canvas (or graph paper, or whiteboard, or cocktail napkin…).

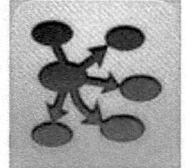

XMind Portable App:

XMind is a cognitive mapping portable app that can be used in a classroom setting. It has a very business-like look to it compared to other apps like it, and it is also customizable.

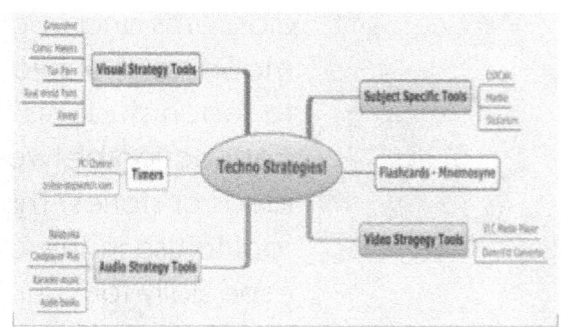

Popplet for iPad:

  Using Popplet, you can use colored pencils and crayons, or even a computer. Students can create a step-by-step map of the story complete with descriptions of events and drawings, or clip art to help them remember what happened.

Smart Diagram Lite:

 Organize ideas and thoughts with clear diagrams. Simple & Easy.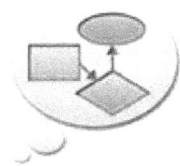

IDEAS FOR USING DIGITAL MIND-MAPPING TOOLS:

1. Teach "concept mapping" to help remember content or take notes.
2. Literary analysis web: similarities and differences of various pieces of literature.
3. Map out the plot.
4. Create a character analysis web.
5. Organize writing (instead of a linear outline).
6. Have students create a cognitive map with words and pictures to help them remember the sequence of events in a story or history lesson.

 **Take it down a level: Apply:** Use "MIND MAPPING APPS" to organize _____ to show _____.

Take it down a level: Apply: Use "MIND MAPPING APPS" to map out what questions students would you ask in an interview with...

To Analyze: Use "MIND MAPPING APPS" to map out the relationship between _____ and _____.

To Analyze: Use "MIND MAPPING APPS" to determine strengths, weaknesses, opportunities and threats present in a story, or real world issue.

Use: "MIND MAPPING APPS" to create a character or plot analysis in English.

Take it up a level: Evaluate: Using graphic organizers, language arts students learn to gather, analyze and evaluate increasingly complex information for writing essays and communicating understanding.

lino - Sticky and Photo Sharing for you

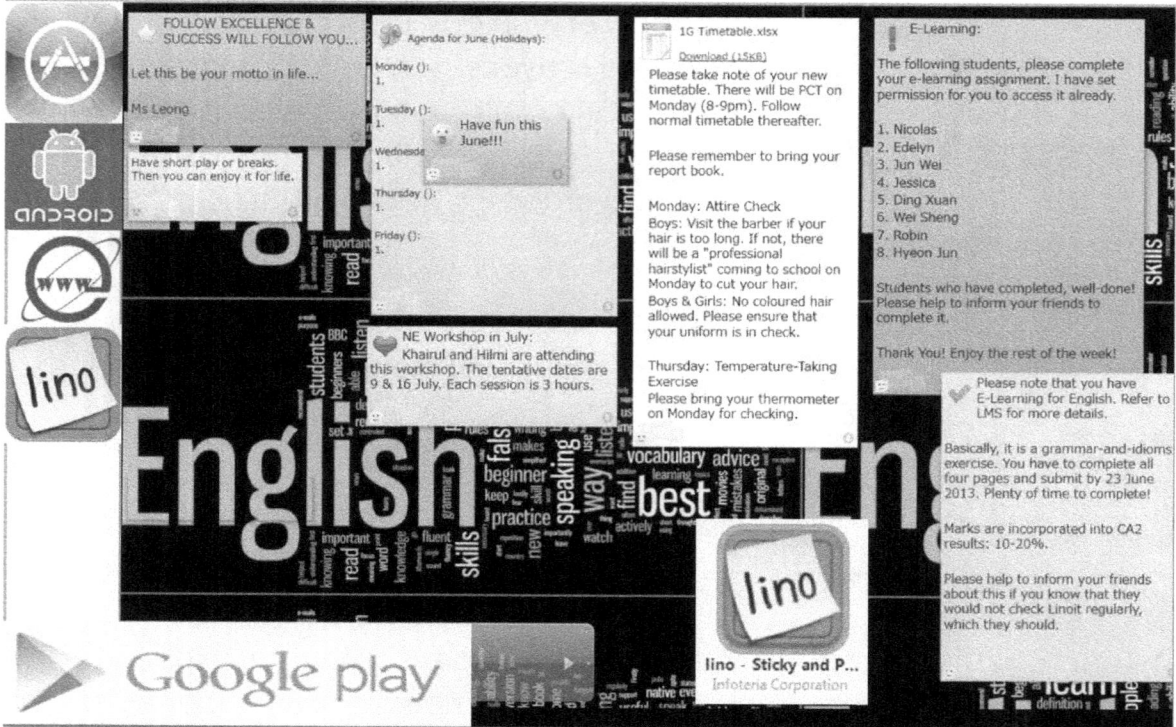

lino is an online stickies service that offers stickies and canvases. You can post, see and peel off stickies on canvases freely. Stickies posted from this iPhone/iPad App can be accessed with PC browsers.

- Take a note right away wherever you are
- Free layout of your pictures and videos
- Share your ideas with your group members

IDEAS FOR USING LINO IN THE ENGLISH CLASSROOM

1. Collecting photographic evidence for analysis.
2. Collaborating to come up with ideas.
3. Collecting words for writing.
4. Reasons for persuasive writing.
5. Can link to Twitter account and fellow Twitter users can add ideas as well.
6. You can customize backgrounds with prompts: writing prompts and questions at a variety of levels of thinking.
7. Students may use lino in the same way that you'd use physical sticky notes to write down their ideas, post them on a wall, categorize them, discuss them and analyze them.

SPEECH TO TEXT: IS IT STILL WRITING?

When most of us think about the skill of writing, we consider writing conventions such as punctuation, spelling, grammar, and other things that make writing consistant and easy to read. We think about sentence structure and paragraph formation. We think about organization. However, a piece of writing can have absolutely perfect writing convention and yet be uninteresting, unimportant, lacking passion, and void of analysis or reflection.

Before technology was an option for putting words to print for people who struggled with writing convention, only those who were able to do both the convention and the creation became authors. While some may believe that is only just, the bigger picture is that many brilliant minds, amazing storytellers, and passionate visionaries were never able to put their ideas into print.

Voice-to-text has changed that reality for those who struggle to put words to paper. Today, doctors, lawyers, business men and women, and authors are using voice-to-text. After voice-to-text captures their stories, their ideas, or their message, proofreaders and copy editors turn the piece into a respectable and publishable document. As a matter of fact, the words you are reading at this very moment are being spoken into a software program called Dragon Naturally Speaking.

In this digital age, it is not only appropriate, but possibly necessary, to teach students how to use these tools to maximize their literary potential.

Siri Dictation App:

 With this dictation app, you can write an email, send a text, search the web, or create a note. With only your voice. Instead of typing, tap the microphone icon on the keyboard and then say what you want to say while your iPad listens.

Like a voiceover app, this app is also a great tool to use for students with vision deficiencies.

"Concerned about privacy? *Why Does The New iPad's Dictation Feature Require A Wi-Fi Connection?*" by Bryan M. Wolfe on Mon March 19th, 2012, discusses privacy concerns expressed by users regarding this app.

http://appadvice.com/appnn/2012/03/why-does-the-new-ipads-dictation-feature-require-a-wi-fi-connection

(This app is built into version 3 of the iPad operating system)

IDEAS FOR HAVING FUN WITH SIRI

Ask SIRI:

1. How much wood would a wood chuck chuck if a wood chuck could chuck wood?
2. Tell me a joke.
3. Knock knock.
4. What do you look like?
5. I need to hide a body.
6. Who's your daddy?
7. Will you marry me?
8. Sing the song.
9. I love you SIRI.
10. What's the answer to the universe?
11. What happened to Hal?
12. Is Santa Claus real?
13. When will pigs fly?
14. What are you wearing?
15. What is the meaning of life? (My favorite)

Dragon Speak:

Paraphrase immediately using Dragon Speak.

Another free strategy to enhance short-term memory so information isn't "gone" in two seconds is to have a student paraphrase what you just taught.

For example, after you've taught something important, ask a volunteer to paraphrase that information for the class. Most likely, your students will not relate the information in the same words you used, which will be novel to the brain.

This strategy only takes seconds to do, yet it lets your students hear the information again, in a different way, with a different voice. The brain likes novelty and, as a result of this strategy, will remember the information better.

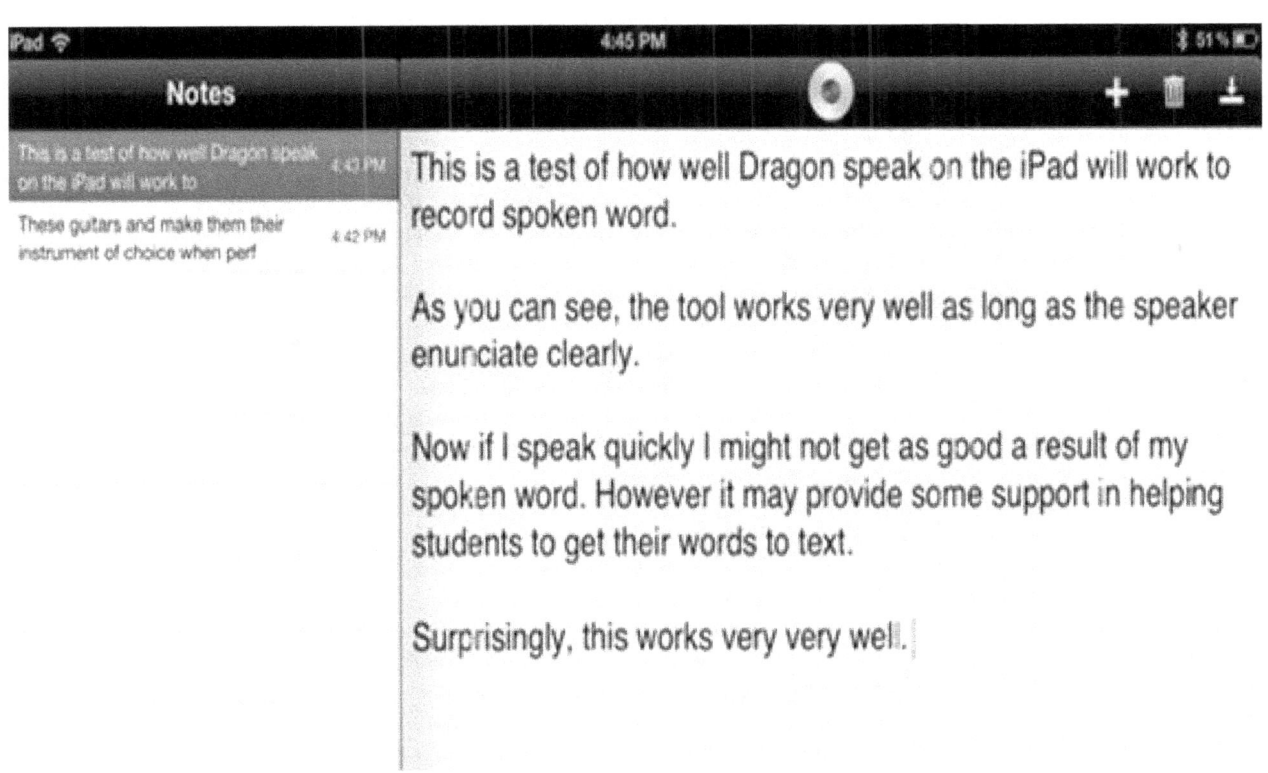

MECHANICS OF WRITING

Dictionary.com

The trusted online dictionary in app form! Provides definitions and synonyms (thesaurus). Also works offline.

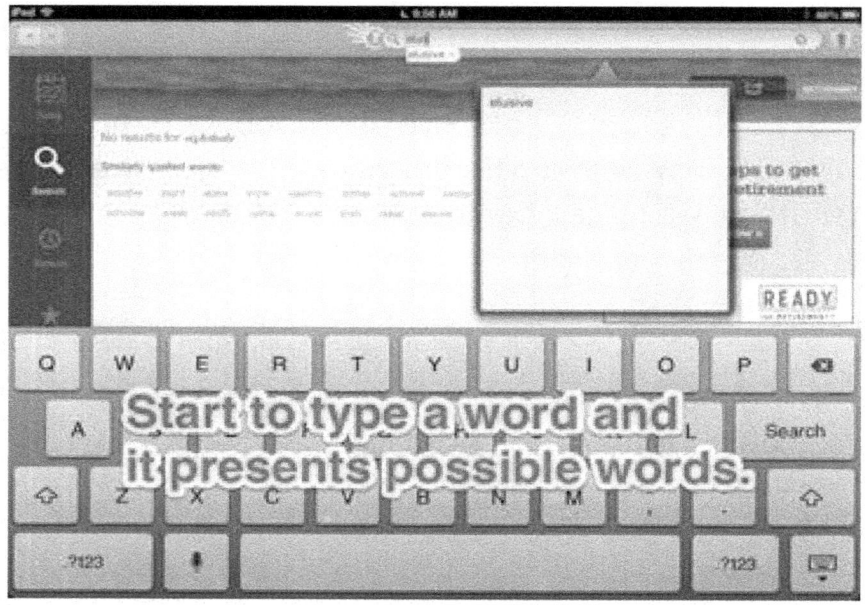

Dictionary.com Flashcards:

The flash card version is originally an iPhone app so it does not present in horizontal mode.

Grammar Up HD:

By Eknath Kadam

Grammar Up is an interactive app that can help learners to improve their grammar, word selection, and vocabulary. Custom timer settings can assist learners to improve response times under exam time constraints. Students are able to study all of the grammar rules by topic.

Note: This app is dry, boring, and no different than using a worksheet or an iPad.

Grammar Up:

By Kuber Tech

Beware: There are two Grammar Up apps. The version by Kuber Tech has higher user ratings than the version by Eknath Kadam. This is the better app.

NOTE: The Grammar UP listed above and the Grammar UP in this row are different applications!

Android Version of Grammar Up By ElmatjLada

NOTE: Check out Grammar Guru

CITATION TOOLS

Easy Bib http://www.breakingnewsEnglish.com/

Son of Citation Machine: http://citationmachine.net/index2.php

Mendeley App: Bibliograpy index cards are no longer efficient or necessary!

Mendeley is Easybig on steroids!

Endnote was the precursor to Mendeley. I'll never forget how thrilled I was when a colleague introduced me to Endnote. Bibliographies and MLA style sheets were the bane of my writing existence. I was able to type bibliographical information into Endnote, or better yet, use an online service to find research and when selecting such research, Endnote automatically captured all the bibliographical information and stored in the database. A plug-in that works with Microsoft Word literally inserted bibliographical information at the click of a button. Mendeley is the modern, digital age, version of Endnote.

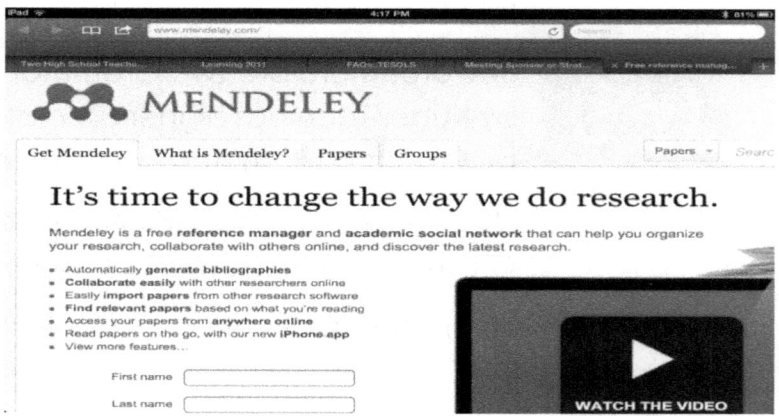

English teachers may cringe at the idea that students no longer need to labor over creating documents in MLA format, or keep their bibliographical information on tediously collected and annotated index cards. However, with tools such as Endnote and Mendeley, that hard work is done for students, thereby allowing them to use their energy in the creative process and focus on higher order critical thinking skills.

If you worry that you are shortchanging your students by allowing them to use these tools because they will need these skills in college, know that college professors are using these tools when writing their research papers. It is more of an injustice to students to deny them the opportunity the digital advantage affords them.

Mendeley is academic software that indexes and organizes all of your PDF documents and research papers into your own personal digital library. It gathers document details from your PDF's, allowing you to effortlessly search, organize, and cite.

MORE WEB RESOURCES FOR THE MECHANICS OF WRITING

CyberEnglish: www.tnellen.com/cybereng/

Writing Lab Online: www.owl.English.purdue.edu/handouts/index2.html

Writer's Resources on the Web: www.writing-world.com/

Guide to Grammar & Writing: ccc.commnet.edu/grammar/index.htm

The Blue Book of Grammar: www.grammarbook.com/

Plagiarism: www.turnitin.com/

Visual Thesaurus: www.visualthesaurus.com/

Thesaurus:

If you've ever been stuck trying to think of another word to describe something, you probably wished you had a thesaurus on hand to help you quickly find the word. This is what the Thesaurus app will do for you.

You will find the right synonym or antonym for just about every word. This is a great tool for both teachers and students to conveniently have with them.

Each search is quick and easy so you don't have to flip through pages in a book to find the word.

Vocabology:

 The makers have an interesting approach to upgrade. Rather than pay a fee to upgrade, you must purchase one of the other products that show up on their bookshelf.

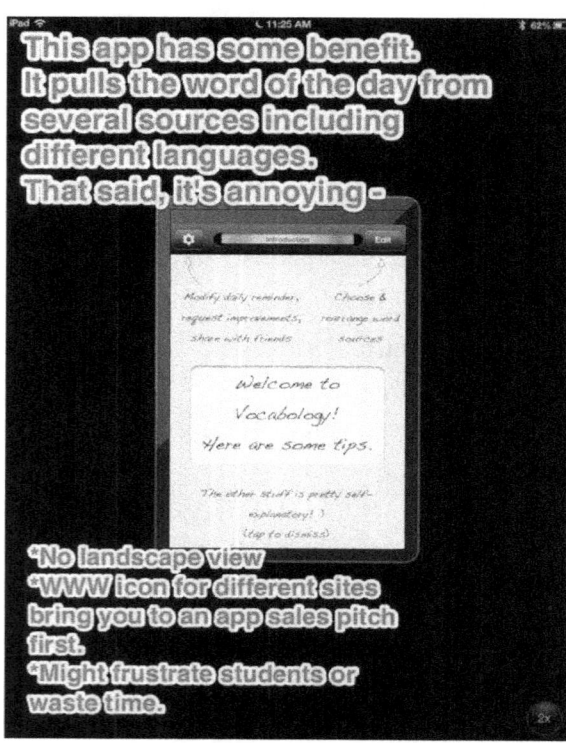

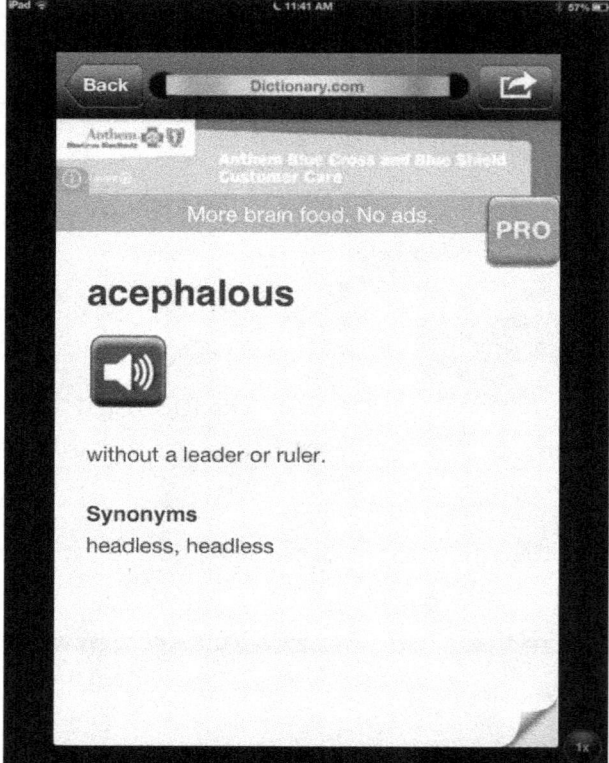

VISUAL CUES - SCIENTIFICALLY KNOWN AS NONLINGUISTIC REPRESENTATION

Children today most often remember what they see better than what they hear, so appealing to a student's visual senses is something valuable to understand when teaching. After all, visual stimulation is paramount to teaching this visually-driven generation. Instead of rejecting it, we must step into their world and embrace it. Then, turn it around to stimulate learning.

THE BRAIN THINKS IN PICTURES:

When learning a subject for the first time, a student might grasp the information better if it's given to them in pictures first and not just words. For example, teach history by showing pictures of former presidents as they looked in their time instead of just telling

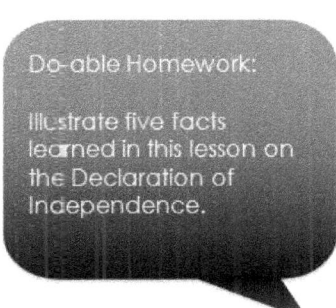

Do-able Homework:

Illustrate five facts learned in this lesson on the Declaration of Independence.

students to read their biographies. Or, teaching new vocabulary words by using cartoons that act out the definitions of the words. The use of visual cues helps to "paint a picture" in the student's mind so that the memory becomes etched in their brain.

TEACHING VOCABULARY WITH VISUALS:

What if we taught English vocabulary in the same way we teach foreign languages? Popular language-learning programs teach vocabulary with visuals – meaningful pictures and other visual cues that the student can easily relate to. Similar strategies can be very effective in the classroom when students are learning words and concepts that are new to them.

Several flashcard packages are available to help teach vocabulary using visuals. Vocabulary Cartoons, from New Monic Books, links vocabulary words with memorable cartoons and captions in order to reinforce understanding and memorization (http://www.vocabularycartoons.com). Philip Geer's, "Picture These SAT Words in a Flash," from Barron's Educational Series (available on Amazon.com), uses a similar approach. These sources both have SAT and ACT vocabulary and Vocabulary Cartoons carries generic vocabulary programs for various grade levels.

I bought their flashcards, sorted through them, and actually found words I had never heard of. One of these was "antediluvian." I love the way it rolls off the tongue. But what does it mean? Antediluvian means prehistoric, ancient, and/or before the flood in the Bible.

The picture on that card is Auntie Lil who is eating dill pickles and reading a book titled "Before the Flood." The flood was ancient, and so is Auntie Lil. Remember this: Auntie, a dill-lovin' lady, is eating pickles and reading a book about very old times. The pronunciation is "auntie-dill-luvian." You can see it in your head – Auntie Dill (pickle) Lovin' Lady. If you flip the card over, you see the definition and some sample sentences using the word.

> *Act out vocabulary words using visual images that will set the word in students' minds forever.*
> *~Fritz Bell*
> *Creative Classrooms*

Picture These SAT Words in a Flash vocabulary cards are somewhat similar. They have a picture on one side and analogies on the other side along with antonyms, synonyms, and sentences.

These catchy visual connections work for many students. You might ask, "But what if our SAT test words aren't on the cards? What if I can't afford to buy them?" Thankfully, these flashcard sets aren't the only way to learn vocabulary with a visual connection.

Rosetta Stone is a computer-based example of a program that successfully builds vocabulary skills combining visuals with auditory feedback and repetition.

VOCABULARY APPS:

Vocabulary impacts test scores, reading comprehension, and fluency. The internet, and the App Store, contain a variety of vocabulary cartoon apps that can be used in the classroom to teach your students new words. The more interesting the cartoon, the more the student will remember the word and even learn to incorporate that word into everyday sentences. This may even encourage them to write stories of their own using their vocabulary words. (This would be a great time to introduce another type of app to teach writing – books and comics.)

MakeBeliefsComix.com:

Comic strips are an outstanding way for students to illustrate their understanding of vocabulary words, demonstrate understanding of a storyline, illustrate a short story by converting it to a comic strip, demonstrate understanding of literary elements, figurative language, and so on and so forth.

Students might use two frames to form pictures to connect to vocabulary for visual vocabulary flash cards.

Try http://www.makebeliefscomix.com, a site with wonderful tools for teachers and students alike.

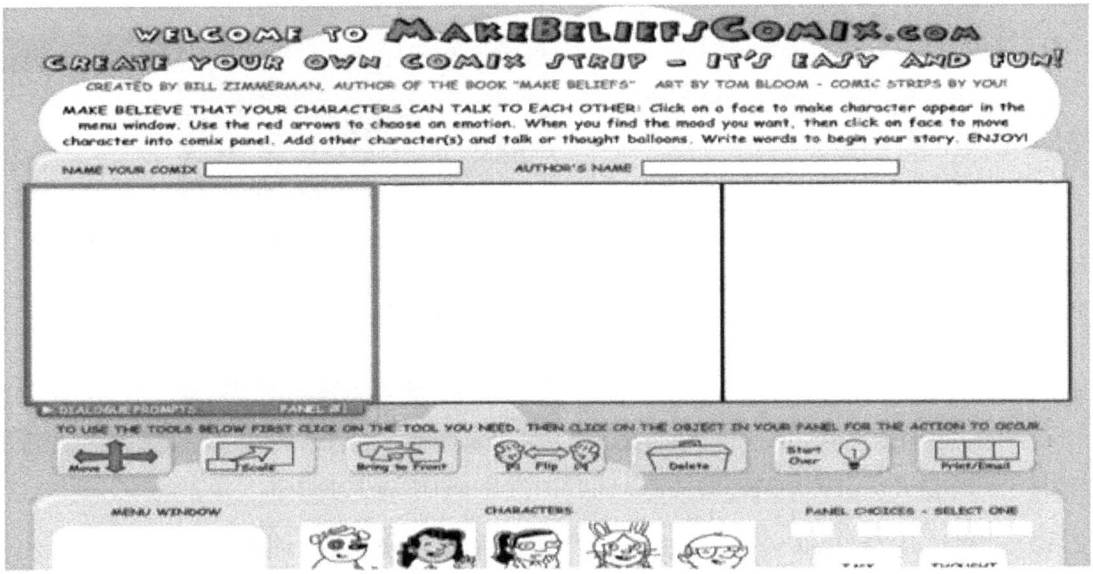

Strip Designer:

 Photos can be added from the camera, your photo-album, or downloaded directly from your Facebook account. You can apply filters to photos and change the layout of the page to fit your needs. You can even paint on the photos or draw your own sketches from scratch.

Use comic strip makers and drawing apps to incorporate technology into research-based practice: Non-linguistic representation.

STORYBOARDS FOR READING COMPREHENSION:

To make a storyboard, without technology, have students fold a piece of paper into squares and draw about what they read. They might do this while they read a story for the first time, as a review with a partner, or for homework after a reading assignment. The process of turning verbal information into a visual format reinforces the learning and helps keep the information in working memory longer.

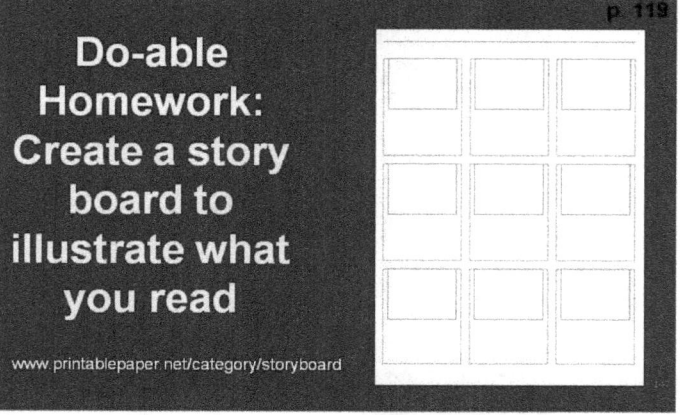

I worked with a teacher who used this strategy with her middle school students. She asked them to draw a storyboard for homework and she gave permission to use the paint program on their computer at home if they wished – so they could either draw it by hand or use the paint program. She was shocked at what came back the next day because her students, whether they used the computer or not, put so much detail into their storyboards. She was clear that they got the information they needed from the reading. She said, "You know, I would never know how well they understood this information if I had assigned a worksheet."

Some students are not as visual as others, so these strategies offer a wonderful opportunity to use peer teaching. Have students pair up and explain their drawings to each other. They can correct or enhance their own drawing while discussing it with their partner.

CREATE A SEQUENCE CHART:

When students need to remember the sequence of events in a story or historical events in order, making a sequence chart for homework is a good option. A sequence chart is like a timeline or storyline and can be drawn on any type of paper, although long adding machine tape works well. Ask students to illustrate with drawings. and describe with words, each event in chronological order.

To remember information in sequence, such as a timeline in history, a cycle in science, or the chronology of a story, use adding machine tape or strips of paper and have students draw their storyboard in sequence. Now they can see the sequence of the storyline, timeline, or process literally in a visual, sequential format. This is a great way to reach kids who struggle with sequences because it reinforces the information with a research-based strategy: non-linguistic representation.

DRAW IT SO YOU KNOW IT – NON-LINGUISTIC REPRESENTATION:

Condense information into a picture and embrace the power of coloring. Teachers often present information verbally and linguistically. However, many of our students are visual learners. A substantial amount of our brainpower is devoted to visual processing. When teachers add a visual component, a drawing component, to what they are teaching, student recall increases.

For example, after teaching for five or six minutes, or up to ten minutes in a high school class, give students three to five minutes to draw a picture, diagram, or symbol of what they just learned. This strategy allows students to take the verbal linguistic information just taught and turn it into visual information. This lets the brain process and use information in a different way which, in turn, helps students to better remember what has been taught.

When we use drawing exercises in the classroom, we often encounter resistance from students. They complain that they can't draw. One way to address this is to draw badly when we draw in the classroom. Use stick figure drawings and emphasize the importance of simple line drawings over drawing well. The point is to create an image that helps us remember what we've learned, not to get graded on our art.

If students say they can't draw, pair them up with someone who doesn't mind drawing. It would be a shame to lose students because of their initial resistance to doing something so different from what they are used to doing in school.

The brain has a huge capacity for visual processing, so the visual component of our memory is very powerful.

Tux Paint:

 This portable app is a computer art software intended for kids ages 3-12. However, there is absolutely no reason why "big kids" can't use it! It is used in schools around the world as a computer literacy drawing

activity. It combines an easy-to-use interface, fun sound effects (that you might want to silence), and an encouraging cartoon mascot who guides students as they use the program. (Tux Paint Config, a separate program, allows parents and teachers to change options – like turn off the sound. There are some accessibility controls.)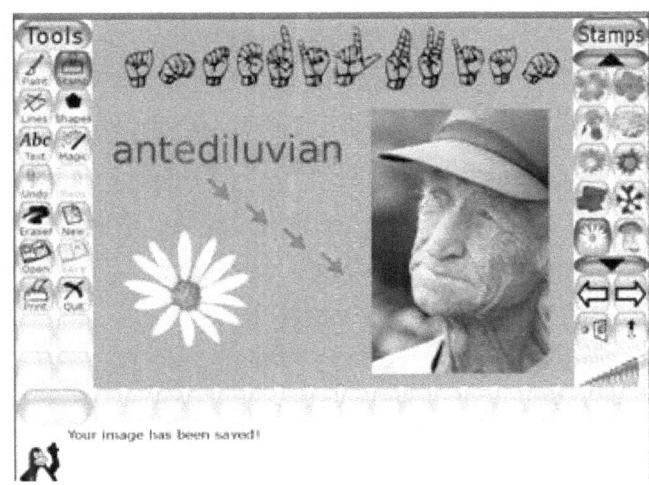

Picasso:

 Draw, Paint, Doodle!

It's simple and intuitive to use, has an almost infinite palette of colors, and has excellent brush effects.

Drawing Pad:

So far, this is the best app I've found for capturing the printed text within pictures. It is a mobile art studio. Students can create their own art using "actual-sized" photo-realistic crayons, markers, paint brushes, colored pencils, stickers, roller pens, and even more drawing supplies.

This is not only useful for art classes, but also for academic classes when students are doing a creative project and they want to use original art to tell a story.

This app is useful for all grade levels to encourage creative fun.

The beautiful user interface puts the fun into creating art!

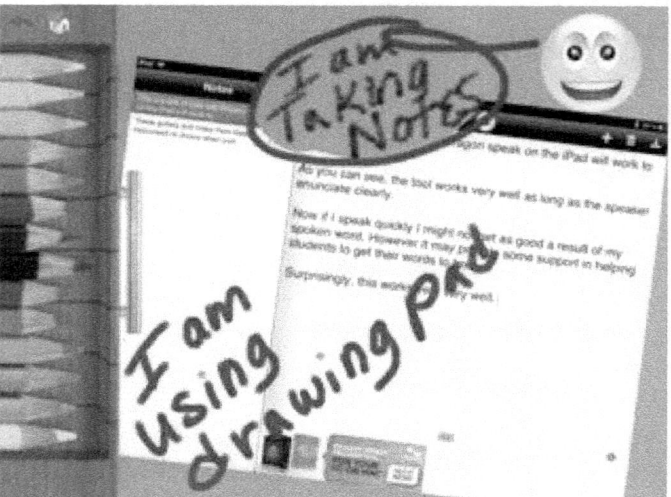

MORE IDEAS FOR USING DIGITAL IMAGE APPS:

1. Draw pictures of the different stages of a story you are reading.
2. Use digital color markers to "colorize" parts of a story or poem (Option: highlight tape.).
3. Create collages to visually represent categories, cause and effect, similarities and differences.
4. Create fliers related to text.
5. Create visual vocabulary flash cards

PhotoCard by Bill Atkinson

By Bill Atkinson Photography

This app allows students to create high quality postcards that can be sent to a recipient via email. They may also be saved as a picture by either using a screen capture program or saving them in a JPEG format.

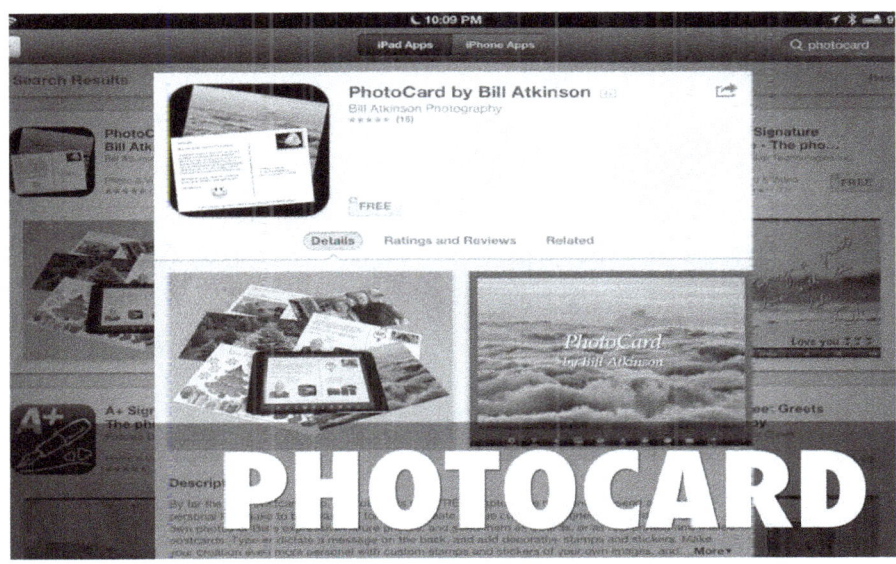

IDEAS FOR USING POSTCARD APPS AS PART OF AN ENGLISH LESSON:

1. Students imagine that they are a character in a story they have read. As that character, they will write home to tell their family about an experience they had, based on the storyline. Students select events from the story that they will detail in their postcard to another character.
2. Students create a postcard as a marketing piece for a travel agency. The setting featured on the postcard depicts a location referenced in their reading
3. Students create a postcard by illustrating a place they would like to visit, and then using 10 adjectives, in a short message, describing the picture in their postcard.
4. Students create a postcard that includes a catchy slogan illustrating a figure of speech.
5. Write a post card to the author sharing what they liked about the book.

*When writing a post card, observe punctuation rules, spelling, and proper sentence structure.

Haiku Deck:

I found an app that has literally transformed how I do slides in my presentations. I'm really aiming for "less is more," as well as a clean look. They make their money by selling additional presentation styles and scenes. However, if you're happy using the choices presented in the free app, you can do a whole lot for no extra cost. I create a slide, then I play the slide until the main menu command disappears from the top of the screen. Then I take a screenshot of my iPad and save the slide to my camera roll. Finally, I upload the slide from my camera roll to my PC using Dropbox . Now I can use the slide in my presentation as a picture. If this sounds complicated, it's not. It's incredibly easy and the results are gorgeous.

IDEAS FOR USING HAIKU DECK AS PART OF AN ENGLISH LESSON:

1. Have students use Haiku Deck to create presentations for Language Arts or English topics.
2. Use Haiku Deck to create visual vocabulary cards.
3. Have students illustrate figures of speech, genre, or other literary elements.
4. Assign a small group to create a presentation depicting the setting and time of a piece of literature.
5. Use to illustrate simile or metaphor.

Toontastic for iPad:

Toontastic is a storytelling and creative learning tool that enables kids to draw, animate, and share their own cartoons with friends and family around the world through simple and fun imaginative play! It's like putting together your own puppet show. The recommended ages for this app is 4-7 and 8-10, however, don't let that stop you from using it as an appropriate app for secondary education. It's not the look of the app that makes it worthwhile or not, it's *how* the app is used!

The number of characters and settings keeps expanding with the addition of themed scene packs – but these packs must be purchased for an additional $.99 each.

Puppet Pals HD:

A simple-looking app with a tremendous amount of potential for secondary application in all subject areas.

IDEAS FOR USING MOVIE MAKERS IN THE ENGLISH/LANGUAGE ARTS CURRICULUM:

1. Students create their own myth/fairy tale or a modern interpretation of an existing myth/fairy tale using digital movie maker software.
2. Students create an original story.
3. Students create a video parody of a news article.
4. Create a book trailer to entice the viewer to read the book.
5. Create a personalized commentary on a piece of literature.
6. Create a public service announcement video on issues addressed in the story and important to a character.
7. Taking on the role of a character in a story, students create a newscast from that character's perspective.
8. Using different genres in literature, students explore and discuss the characters from the current lesson's reading material. With an understanding of those characters, students create a video depicting the characters in a modern day dilemma.

Eyejot Video Mail

Eyejot is a video mail solution for mobile devices. With Eyejot, students can easily create and send video messages to anyone in the world - whether or not they're using Eyejot. Compose new videos using iPhone or iPod Touch or select existing video from your photo gallery. Easily send a single video message to one or many people.
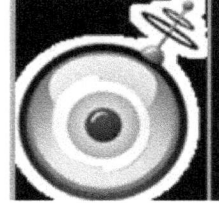

IDEAS FOR USING EYEJOT AND OTHER VIDEO APPS

1. Consider having students use Eyejot to quickly summarize new material and then email it to a peer.
2. Have students explain how they might use a concept they've learned in 60 seconds or less.
3. Explain the author's intent _____ and then email their response via video to the teacher, another classmate, or peer (consider peer tutoring).

GarageBand:

I had not seen the academic value of GarageBand, with the exception of using it in a music class or program, until I saw how a teacher had her students create their own background music for project videos instead of using ready-made music they might find online. This avoided any copyright infringement, as well as challenged students to use their imagination and skill to create unique background music.

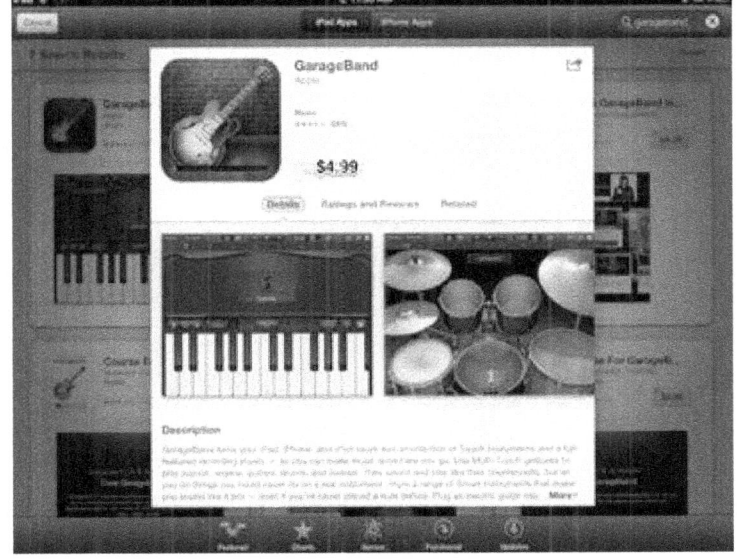

This also had the advantage of giving the musically inclined students a way to contribute to a content subject area project.

Students would have to use their analytical skills, or trial and error, or research skills, to figure out how to use GarageBand to create the music. There is academic benefit in that exercise.

Apps that fit into the "evaluating" stage improve the user's ability to judge material or methods based on criteria set by themselves or external sources. They help students judge content reliability, accuracy, quality, effectiveness, and reach informed decisions.

Does the app help the user...

- Check for accuracy?
- Detect inconsistencies?
- Monitor effectiveness?
- Evaluate procedures?
- Critique solutions?
- Appraise efficiency?
- Judge techniques?
- Contrast performance?
- Check the probability of results?

Side by Side

 The iPad does not lend itself to comparing two websites side by side. It's mainly a one app at a time device. Side by Side allows students to view multiple browser windows side by side. Students can practice comparing and contrasting, using detective skills to evaluate differences, benefits, deficiencies, etc.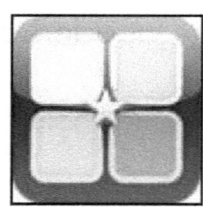

inDecision

Use inDecision when determining the positive (pro) and negative (con) features of a technique, function or decision. List aspects of each side on the T-frame and then rate its level of importance. The combined results of all the data points automatically transforms into a bar graph with percentage figures. Students will be able to evaluate whether or not possible solutions will meet the desired outcome.

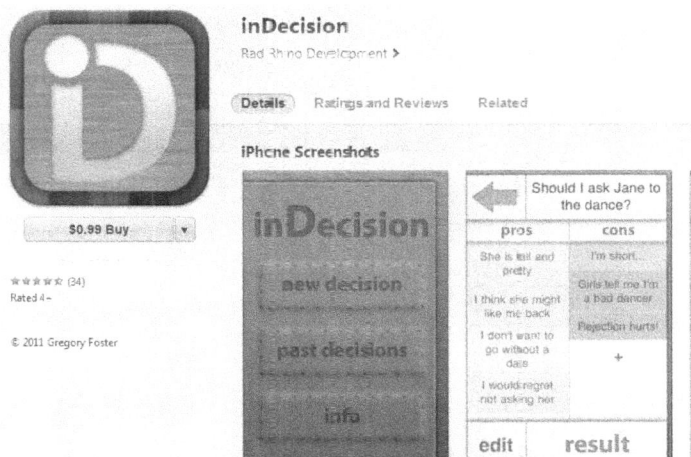

Edmodo:

This app makes it easy for teachers and students to stay connected and share information in one localized space. Use your iOS device to send notes, submit assignments, post replies, and check messages and upcoming events while away from the classroom.

Teachers can post last-minute alerts to their students, keep tabs on recent assignment submissions, and grade assignments. Students can view and turn in assignments and check their latest grades. Class discussions can be conducted securely, both during and outside of school hours.

Learnist

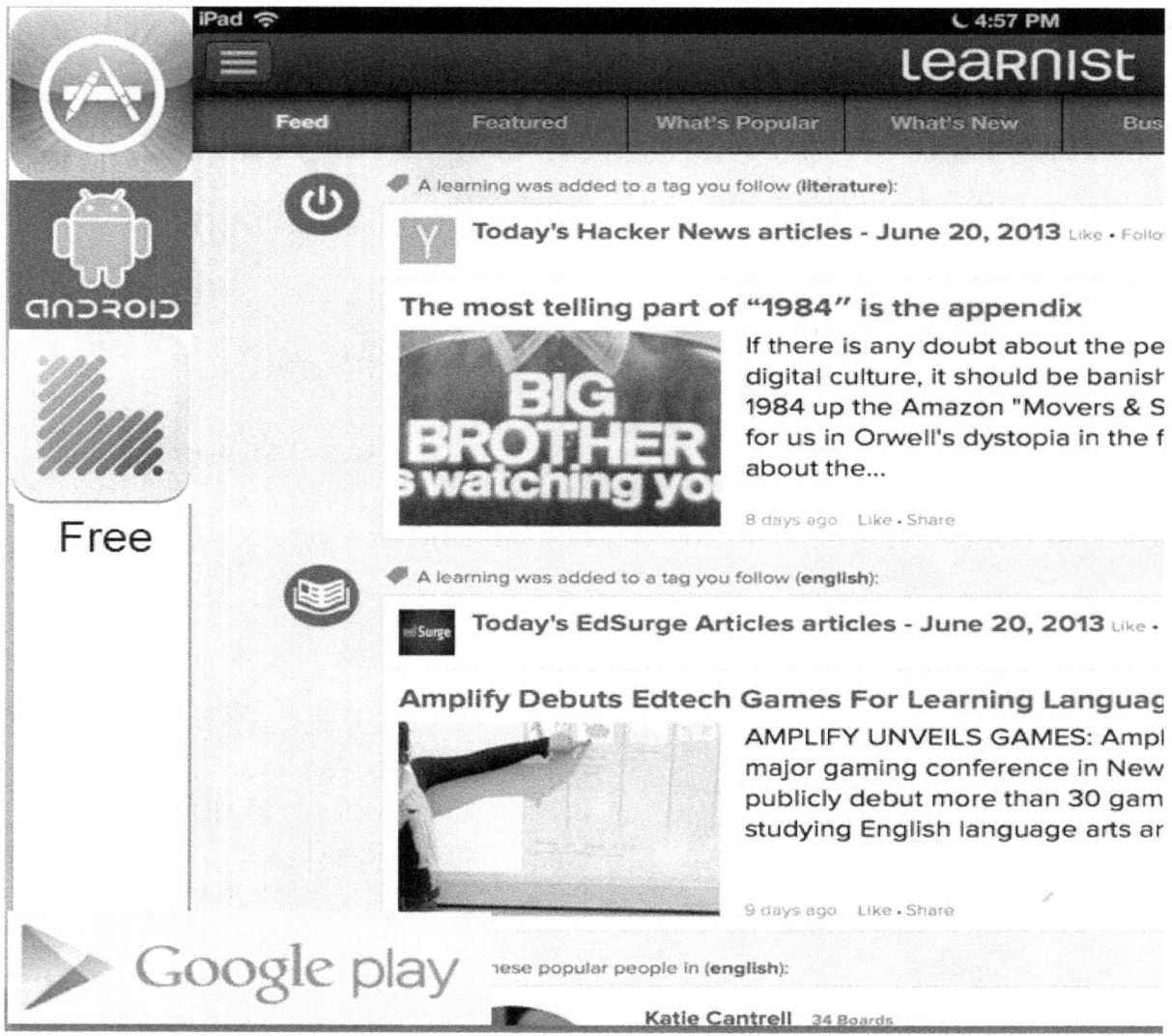

Evernote on Steroids

"By introducing it at a young age, teachers are able to develop the habits of the mind that are essential for students to be good digital citizens. Students learn to use appropriate language, to speak kindly and with compassion, to be supportive rather than critical, and to ask thoughtful questions. ~ C.M. Rubin., Education World

*Evernote – see ORGANIZING RESEARCH AND DIGITAL FILES in the TOC for more information about Evernote.

ShowMe Interactive Whiteboard

ShowMe allows you to record voice-over whiteboard tutorials and share them online.

This app is great for building tutorials. Students can make their own Khan Academy-style videos and post them online.

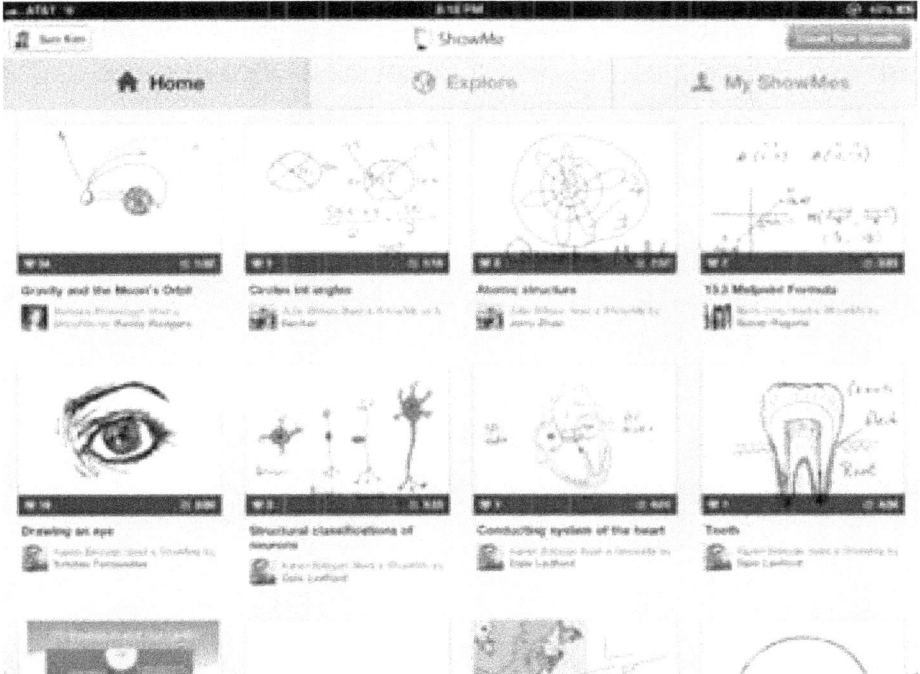

Android: Whiteboard by Henry Huang

Educreations Interactive Whiteboard

Educreations virtual whiteboard can be used by classroom teachers or their students to create presentations, mini-lessons, or to document learning. It may be used either online through a browser or with an iPad, providing a versatile platform for student to apply what they've learned.

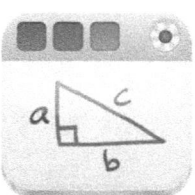

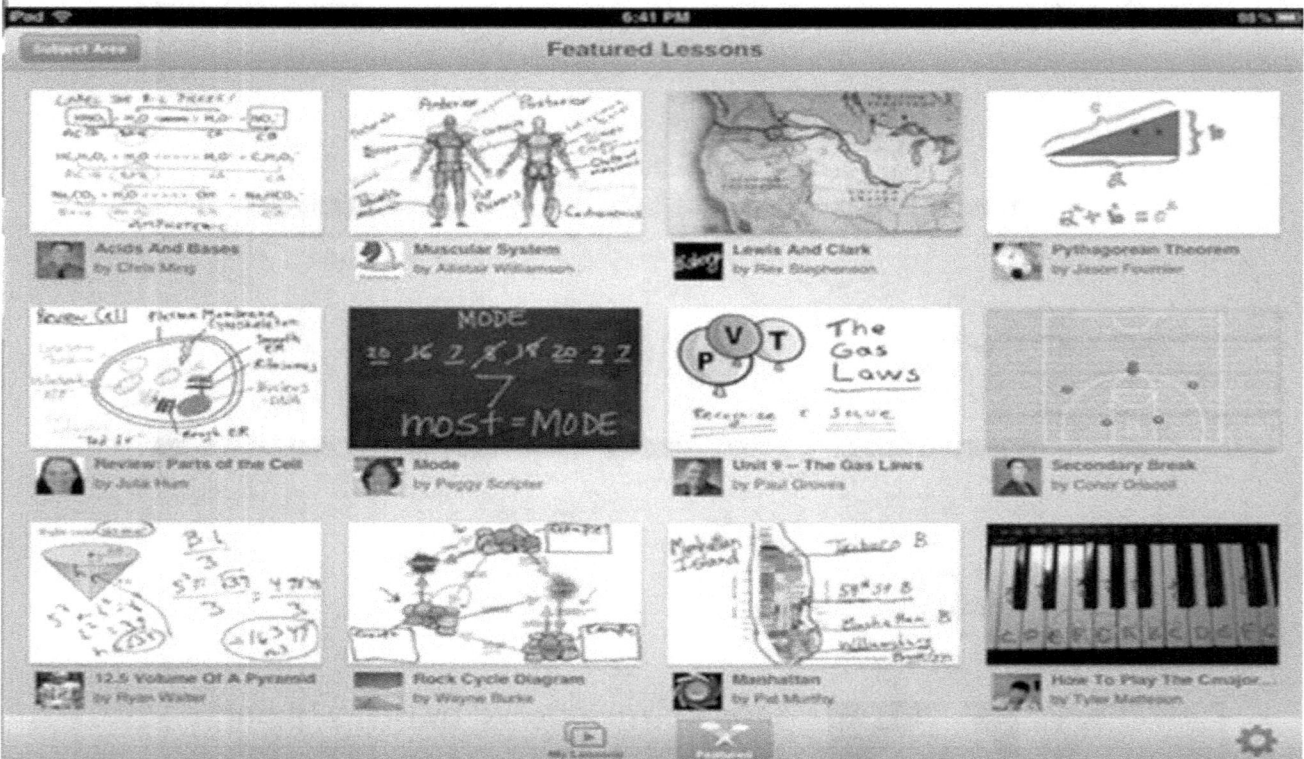

IDEAS FOR USING EDUCREATIONS:

Ask students to use Educreation to respond to the following questions:

- How would you show your understanding of _____ ?
- Show what elements you might change to _____.
- What facts would you select to show _____?
- Illustrate your plan to _____.

Take it up a level: Analyze: Use EDUCREATIONS to show the parts or features of _____.

What is the function of _____?

Explain Everything

Explain Everything is the first truly user friendly interactive white board I've found for the iPad. If you could only have one whiteboard app, this is the one I'd recommend even though it is not a free app. What are the benefits? First it's incredibly flexible.

Teachers can

- import PowerPoint presentations, PDF files, and other documents.
- add animation.
- annotate and narrate the documents imported into the app creating a teacher made tutorial.
- share teacher made creations with students for whole class instruction or individualized instruction, as well as flipped classroom.
- record Lessons.
- create demonstrations.
- export videos to a teacher or student blog to support instruction, or directly to Edmodo.

Students can

- create their own video tutorials for what they've learned in the classroom.
- reinforce lessons taught in the classroom outside of class.
- create videos that can be used for assessment - and it's FUN!

RESEARCH APPS

Wolfram Alpha:

> "Wolfram Alpha is a computational knowledge engine, not a search engine. It works by using its vast store of expert-level knowledge and algorithms to automatically answer questions, do analysis, and generate reports. Wolfram Alpha's long-term goal is to make all the world's knowledge systematically computable and accessible."
> http://www.wolframalpha.com/

Wolfram Alpha appears to be a valid tool for the math and sciences, however, it also has a place in Language Arts. Compare texts and include literary statistics. Use Wolfram Alpha to determine the number of lines attributed to a character in a play thereby being a useful tool in assigning roles to non-readers, introverts, extroverts, and animated, uninhibited personalities!

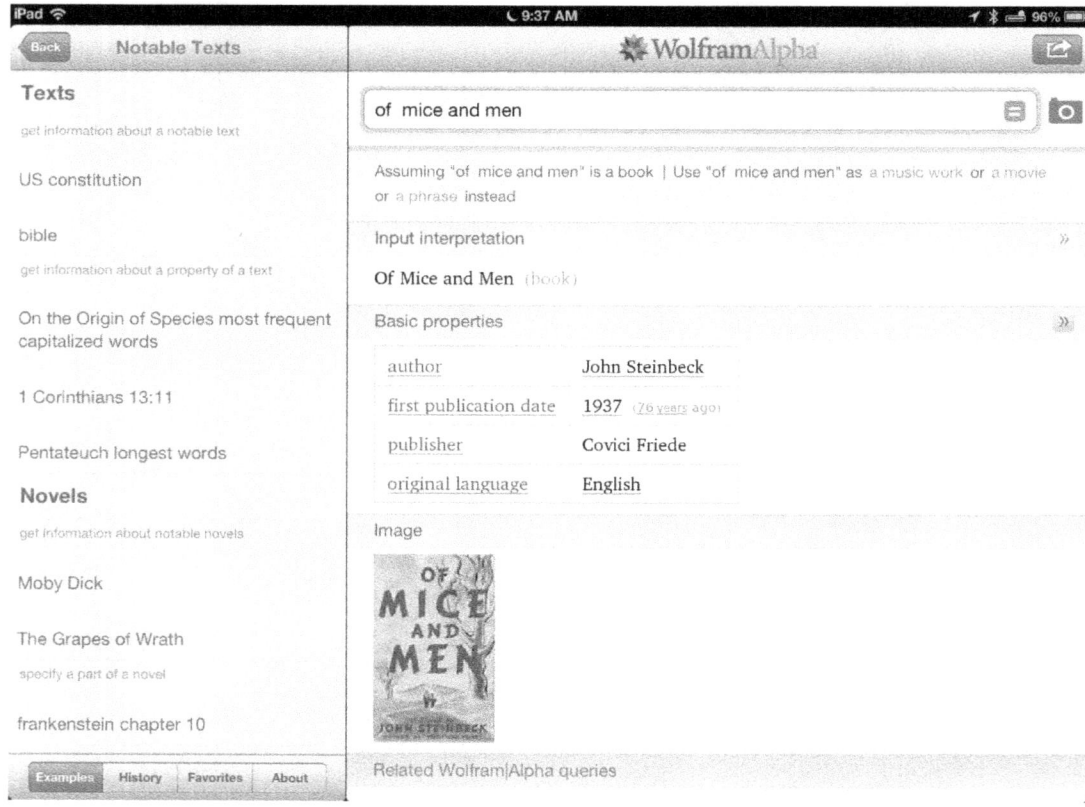

A quote from my 22 year old son, who is in Engineering School. "Best app ever! May have to pay to do the more complex fun stuff, but it's awesome and you can always go online and use it there too if you don't want to use the app."

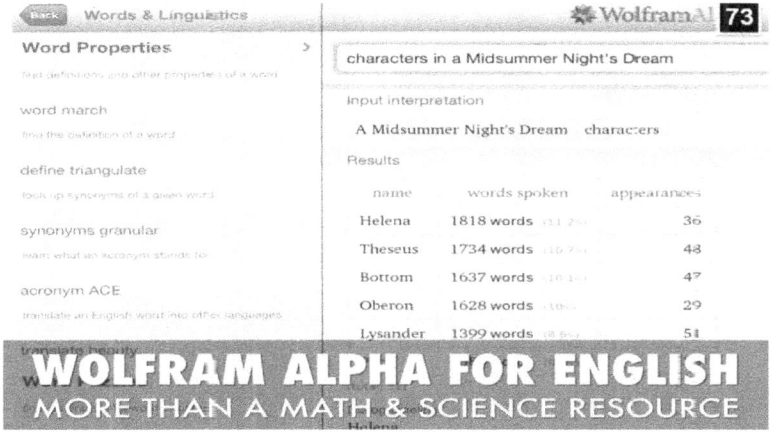

IDEAS FOR USING WOLFRAM ALPHA IN ENGLISH/LANGUAGE ARTS:

1. Use Wolfram Alpha to assign roles in a play.

   When dramatizing plays in the classroom, use Wolfram Alpha to find out the characters in the play, the number of appearances they make and how much they speak.

   With that information, you can strategically assign parts to students in the class by matching their personalities and skills with the characters.

   For example, because Helena appears in the play 36 times and speaks 1818 words, you would know to give that part to a student who loves to be the star of the play, reads well, with emotion, animation, and strength of voice. You would know to avoid giving that part to a student who is extremely shy and reads barely above a whisper.

2. Compare & contrast

   When assigning a research paper comparing pieces of literature, add a statistical and data component by using Wolfram Alpha to provide comparison data between texts.

   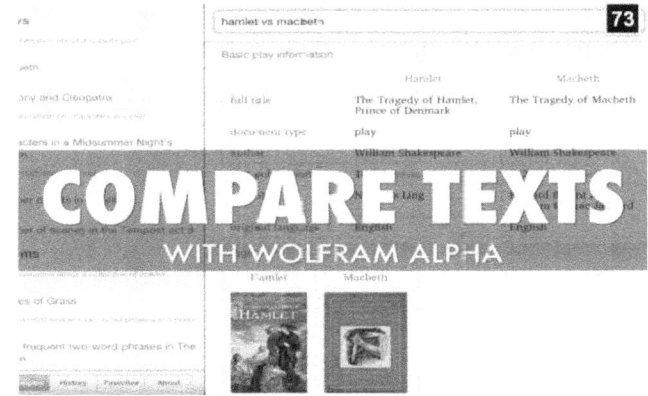

   Ask students what the data might tell them about the differences and similarities in the texts.

How Stuff Works:

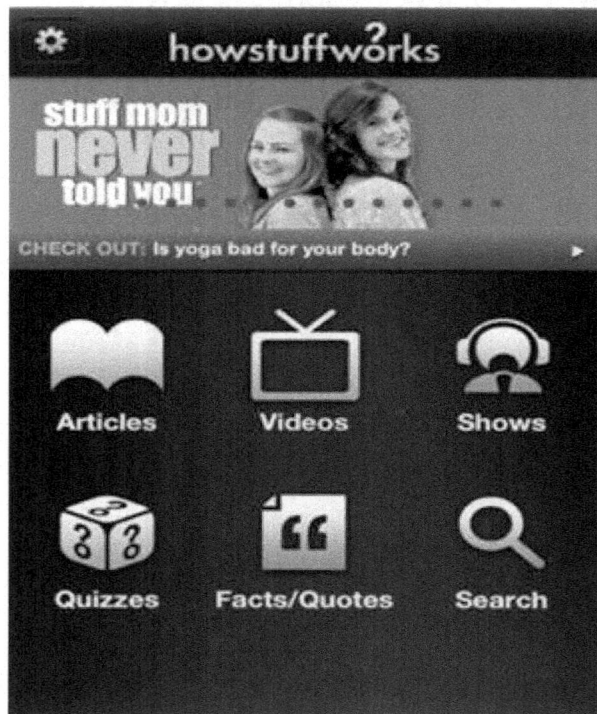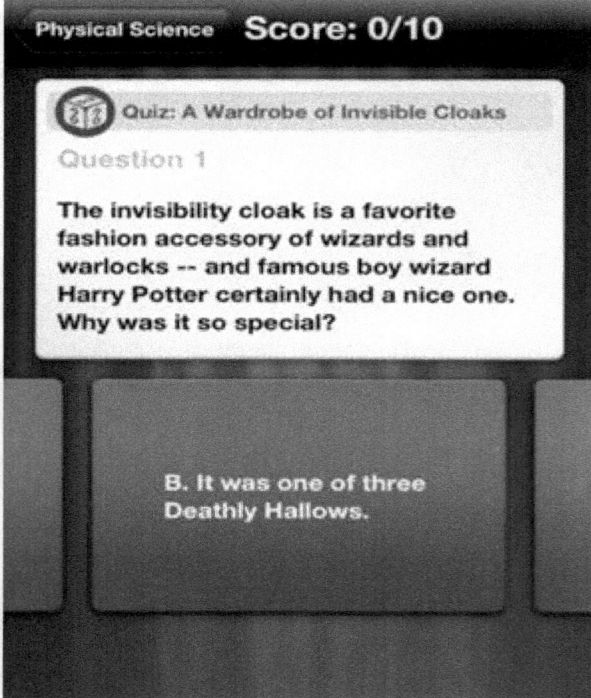

40,000+ articles! 12,000+ videos! 2,000+ shows! 1,000+ quizzes! Students can tweet their favorite podcasters while listening to shows that provoke thought and foster critical thinking on topics relevant to academic curricula and standards. Research objects, activities, machines, etc. referenced in texts fostering deeper understanding of the written word.

HowStuffWorks includes 30,000+ articles. It's possible to watch archived videos from both HowStuffWorks and the Discovery Channel!

CLOUD STORAGE AND TEACHER TOOLS

There are a variety of apps that are created just for teachers that help to make our jobs more efficient when incorporating iPads into our instruction.

Dropbox:

Dropbox is a free service that lets you access your photos, documents, and videos anywhere. Any file you save to your Dropbox is accessible from all of your computers, iPhone, iPad and even the Dropbox website!

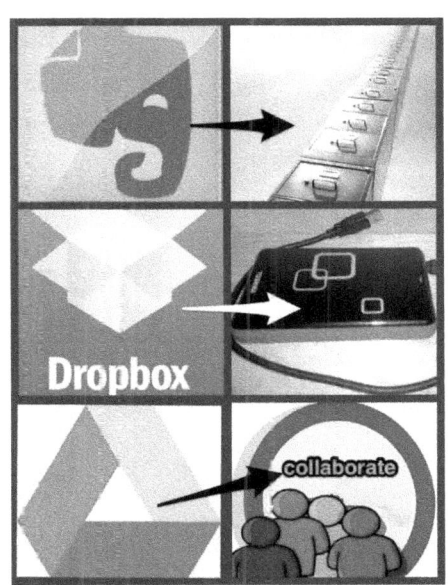

With this accessibility app you can take your work on the go and review docs when you're out and about.

Dropbox can be used by students to view/download their textbooks and other course content.

The free version of Dropbox includes 2GB of file storage, which should be more than enough for most classes or courses. If you exceed the free storage limit, then you'll need to upgrade to a paid account. It is also important to note that shared files and folders count toward all users' storage limits. So plan what you share, and who you share it with, carefully.

Note: Just as any file you add will be accessible from all of your computers, any file you delete will be deleted from all of your devices. It is not a good idea to "drag and drop" files when working with Dropbox. Instead, always copy and paste your files.

File Conversion Website that's Free and Actually Works!

Don't spend another dime on conversion software. CloudConvert will convert files for free! https://cloudconvert.com/

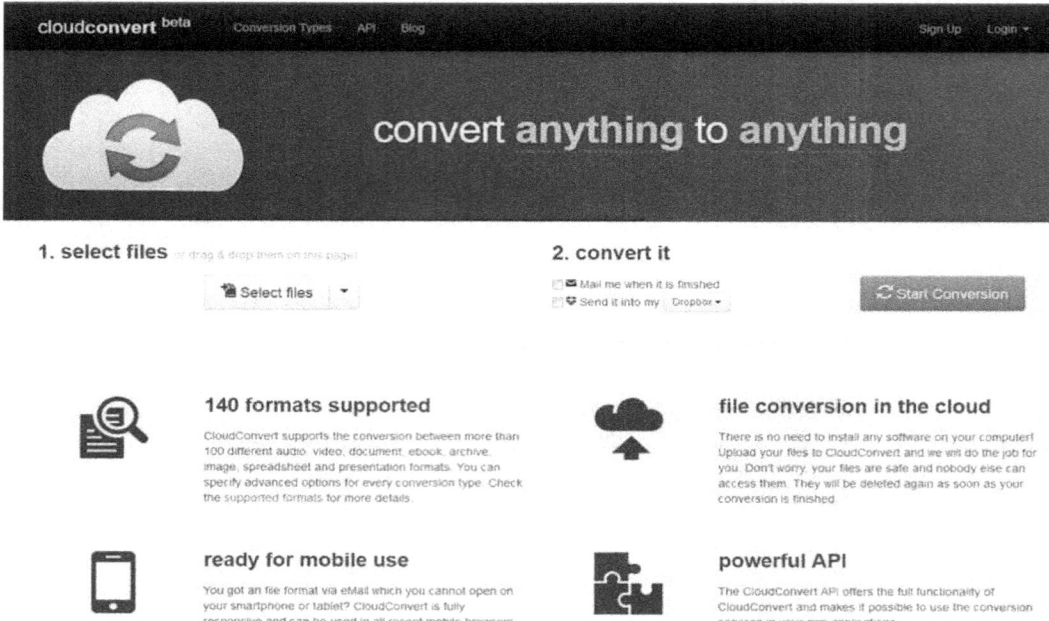

Text Compactor – Remember AutoSummary Tools in Microsoft Word?

Create your own version of Spark Notes with Text Compactor. Just copy and paste a long document into the app and follow the prompts. The redundancy and unnecessary content is removed leaving the reader with a shortened version of the text! ***Check the shortened document to ensure that important information is not removed. The app is not perfect, but it's pretty darned good!

HAVE FUN!

I'm sure that, by the time you read this book, there will be countless additional apps for the iPad, Droid, and Windows tablets. The one certain thing about technology is: it changes and it changes fast.

Exploring the different options available, sorting out those worth your, and your students', time and those that are junk can be time consuming. Learning new apps also requires time - some more than others. However, it truly is worth the effort when you find a piece of technology that increases productivity, achievement, or is simply brain engaging and fun.

If you are a technophobe, I encourage you to step out of your comfort zone and explore. Who knows what you might discover that would add value to your tech experience and life?

I had never touched an iPad before I accepted the challenge to write a book on the topic and develop a technology seminar. Though I had some very frustrating moments (and occasional hours), I've found many apps to be extremely beneficial and the effort to be well worth the time.

Dive in!

DON'T FORGET TO DOWNLOAD THE SUPPLIEMENTAL FREEBIES!

For access to your

FREE

collection of supplementary materials

specific to this text and seminar,

go to tech.susanfitzell.com

A PDF of the corresponding Seminar Presentation with several additional apps(poetry, writing prompts, book creation apps) is included in the free download!

Bright Ideas

Bright Ideas

www.ingramcontent.com/pod-product-compliance
Lightning Source LLC
Chambersburg PA
CBHW081842170426
43199CB00017B/2815